KEY MATHS 9

First published in 1997 by:
Stanley Thornes (Publishers) Ltd

Second edition published in 2001 by:
Nelson Thornes Ltd
Delta Place
27 Bath Road
CHELTENHAM
GL53 7TH
United Kingdom

03   04   05   /   10   9   8   7   6   5   4   3

A catalogue record for this book is available from the British Library.

ISBN 0 7487 5987 5

Illustrations by Maltings Partnership, Hugh Neill, David Oliver, Angela Lumley, Jean de Lemos, Clinton Banbury
Page make-up by Tech Set Ltd

Printed and bound in China by Midas Printing International Ltd

Acknowledgements
The publishers thank the following for permission to reproduce copyright material:
Allsport: 214 (JohnGichigi), 132, 134 (Mike Powell), 64, 215 (Simon Brut); 216 (Steve Dunn), 213 (Mike Cooper); Martyn Chillmaid: 9, 10, 12, 26, 28, 29, 31, 33, 42, 43, 46, 47, 48, 60, 61, 70, 77, 93, 95, 110, 115, 116, 123, 126, 130, 148, 161, 162, 166, 167, 168, 180, 186, 193, 228, 278, 280, 291, 300, 303, 304, 306, 309, 310, 313, 352; Collections: 154 (Neil Calladine); Colorsport: 133 (Andrew Cowie), 134; Empics: 136, (Claire Macintosh); 221; John Birdsall Photography: 101, 106; NASA: 204; PPL Limited: 1; P&O Ferries: 350, 353, 359; Robert Harding Picture Library: 111; Rex Features: 149, 201; Science Photolibrary: 236; Sylvia Cordaiy Photo Library: 235 (Patrick Partington); Telegraph: 114 (Fred Ward/Black Star); Tony Stone Images: 102; Topham Picture Point: 350; Travel Ink: 25, (D Toase); All other photographs STP Archive

Cover photographs: Tony Stone Images (front);
Pictor International (spine); Tony Stone Images (back)

The publishers have made every effort to contact copyright holders but apologise if any have been overlooked.

KEY MATHS 9¹

▶ **David Baker**
The Anthony Gell School, Wirksworth

▶ **Paul Hogan**
Fulwood High School, Preston

▶ **Barbara Job**
Christleton County High School, Chester

▶ **Renie Verity**
Pensby High School for Girls, Heswall

Contents

 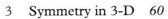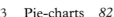
iv

1 Area

The *Cutty Sark* is an old merchant sailing ship which is kept at Greenwich on the River Thames. It was built in 1869 with a total area of sails of 32 000 square feet or 2972 square metres.

CORE

1 Counting shapes

Malcolm is going to change the
shape of his patio.
He wants it to be the same area.
He must use all the slabs in his
new design.

Malcolm has left spaces for plants
in his new design.

Exercise 1:1

1   Use squared paper for this question.
    a Draw a rectangular design for the patio.
        Remember you must use all 16 slabs.
    b Make some designs of other shapes for the patio.
        You may leave some spaces for plants in your designs.

Area	**Area** is measured using squares.

1 cm² is the area of a
square 1 cm by 1 cm.

1 cm

1 cm

This shape has 5 squares.
The area of this shape is 5 cm².

2 These shapes are made with centimetre squares.
Find the area of each shape.

a   **b**   **c**

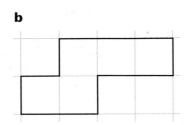

3 Veena has made these shapes with triangles.
Two triangles put together make 1 cm².
Find the area of each shape.

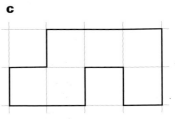

a   **b**   **c**

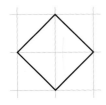
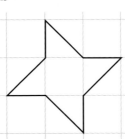
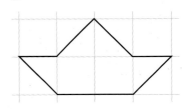

4 These shapes are drawn on axes.
Find the area of each shape.

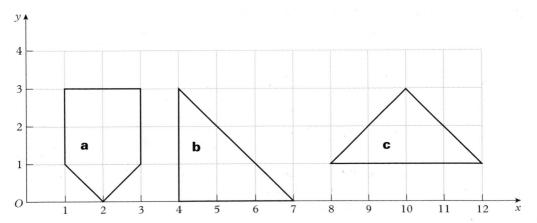

5 Look at this list of items:
dinner plate, table cloth, classroom floor, field, stamp, hand.
  a Which has the smallest area?
  b Which has the largest area?
  c List the items in order of area.
  Start with the smallest.

Some shapes do not fit on to grids exactly.
You can only estimate the area.

(1) Count the whole squares.
(2) Count the squares that are more than half full.
(3) Add them together to make your estimate.

Example   What is the area
             of this pond?

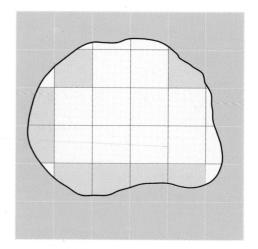

You can only **estimate** the area.
Count whole squares first.
There are 10 of these.

Now count squares which lie
more than half inside the outline.
There are 8 of these.

An estimate of the area is
10 + 8 = 18 squares.

Exercise 1:2

Lorna works at a garden centre.
She tells people how many fish can live in each pond.
Two fish can live in each square of pond.
The pond in the example above has an area of about 18 squares.
So 18 × 2 = 36 fish can live in this pond.

The diagrams show six ponds.
a Estimate the area of each pond.
b Work out the number of fish that can live in each pond.

1
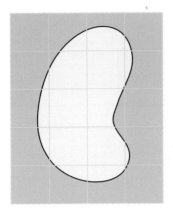

4
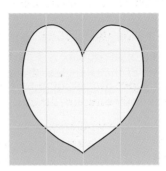

2
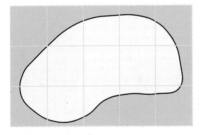

5
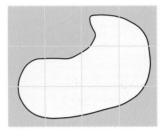

3
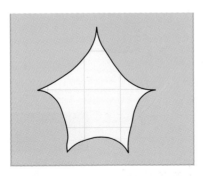

6
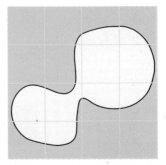

Pat has stuck two shapes together.

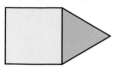

The area of the square is 4 cm².
The area of the triangle is 2 cm².
The area of Pat's new shape = 4 + 2
                            = 6 cm²

Exercise 1:3

1 Pat has made some other shapes.
Find the area of each one.

a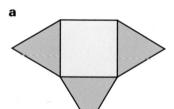

b

c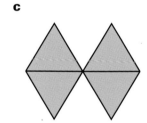

2 The rectangle and the square shown below are not drawn to scale.
The area of the rectangle is 7 cm².   The area of the square is 4 cm².

Find the area of each shape.

a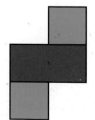

b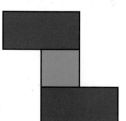

c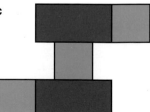

3 The area of this rectangle is 12 cm².   The area of this rectangle is 8 cm².

Find the area of each shape.

a 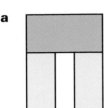   **b**   **c**

4 The area of this shape is 10 cm².   The area of this shape is 7 cm².

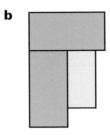   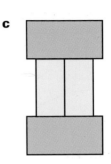

Find the area of each shape.

a    **b**    **c**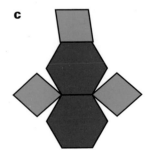

● **5** Carl is making shapes using rectangles and triangles.
The area of a rectangle is 16 cm².
Each triangle is half the area of a rectangle.
Find the area of Carl's shape.

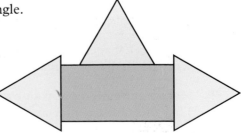

▼ *Exercise 1:4*

Cara has drawn these shapes.
Each shape is labelled with its area.

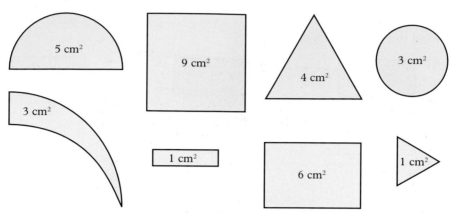

5 cm²   9 cm²   4 cm²   3 cm²

3 cm²   1 cm²   6 cm²   1 cm²

Cara cut out her shapes and made these patterns.
She wrote the total area of each pattern underneath.

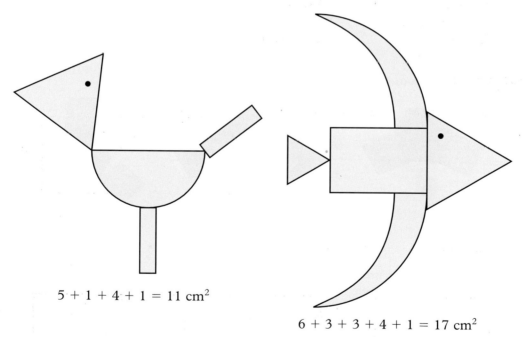

$5 + 1 + 4 + 1 = 11$ cm²

$6 + 3 + 3 + 4 + 1 = 17$ cm²

1   Use Cara's shapes to make some patterns of your own.
Write down the total area of each pattern.

2   Make some patterns of your own.
They must have a total area of:
a 24 cm²   **b** 20 cm²

2 Areas of rectangles and parallelograms

Each row on this sheet has 5 stamps.
There are 2 rows.
There are $2 \times 5 = 10$ stamps altogether.

Exercise 1:5

1 **a** How many squares are there in
this rectangle?
b What is the area of the rectangle?

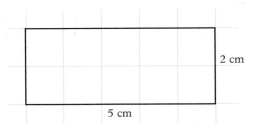

2 cm

5 cm

2 Find the area of each rectangle.

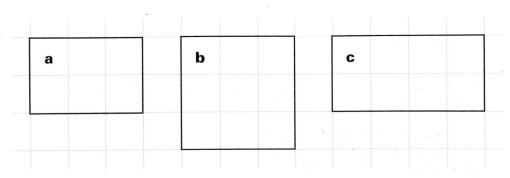

a   b   c

Area of a rectangle	**Area of a rectangle** = *l*ength × *w*idth = $l \times w$	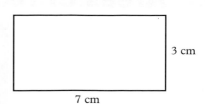 3 cm 7 cm

Example

Calculate the area of this rectangle.

Area = length × width
= 7 × 3
= 21 cm²

Find the areas of these rectangles.

3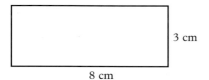

3 cm

8 cm

5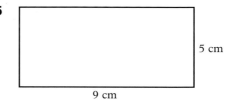

5 cm

9 cm

4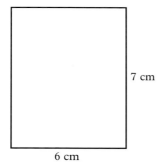

7 cm

6 cm

6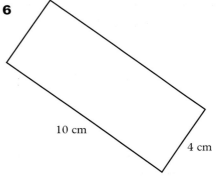

10 cm

4 cm

7 What is the area of this playing card?

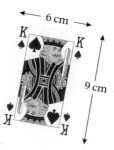

6 cm

9 cm

8 What is the area of this CD case?

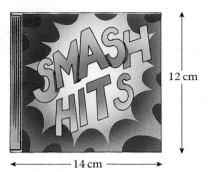

12 cm

14 cm

Area can also be measured in m².

1 m² is the area of a
square 1 m by 1 m.

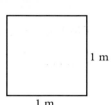

1 m

1 m

---

9 A swimming pool is 15 m long and 4 m wide.
What is the area of the pool?

10 A playing field is 250 m long and 175 m wide.
What is the area of the playing field?

● **11** Kay has drawn a rectangle and a square.
They are both the same area.

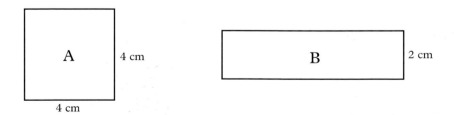

A

4 cm

4 cm

B

2 cm

a What is the area of square A?
b What is the length of rectangle B?

● **12** This box of cornflakes has length 23 cm, width 7 cm and height 30 cm.

   a Find the area of the front of the box.
   b What part of the box has the same area as the front?
   c Find the area of the top.
   d What part of the box has the same area as the top?
   e Find the area of the side of the box.
   f Find the total area of all the faces of the box.

Remember: The faces are the top, the bottom, the front, the back and the two sides.

. . . .

| **Area of a parallelogram** | A parallelogram can be changed to a rectangle by moving the triangle. |

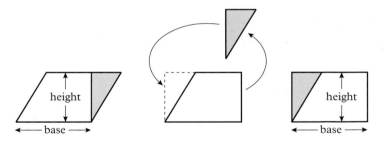

Area of a parallelogram = Area of a rectangle
= base × height

Example

Find the area of this parallelogram.

Area = base × height
    = 7 × 4
    = 28 cm²

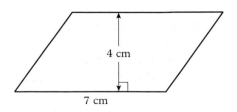

4 cm

7 cm

Exercise 1:6

Find the areas of these parallelograms.

1

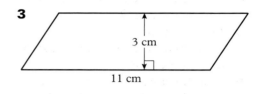

8 cm

10 cm

3

3 cm

11 cm

2

12 cm

5 cm

4

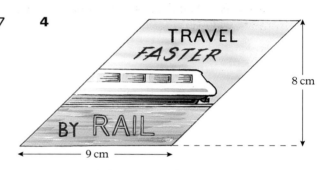

TRAVEL FASTER

BY RAIL

8 cm

9 cm

5 David has drawn a parallelogram.
The area is 40 cm².
The base is 10 cm.
What is the height of David's
parallelogram?

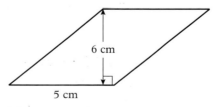

height

10 cm

6 These two shapes have the same
area.
  a What is the area of the
  parallelogram?
  b What is the area of the
  rectangle?
  c What is the length of the
  rectangle?

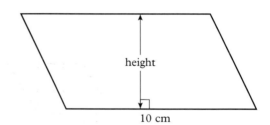

6 cm

5 cm

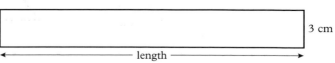

3 cm

length

Exercise 1:7

1 John paints small pictures of animals.
He sells them at different prices:

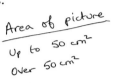

Area of picture	Charge
Up to 50 cm²	£5
Over 50 cm²	£8

Find the area of each of these pictures.
Write down how much each picture would cost.

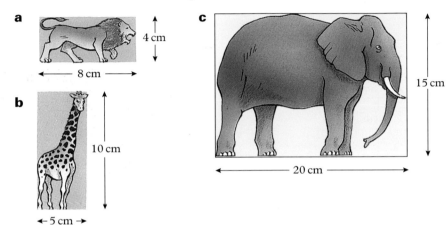

a   4 cm   8 cm

b   10 cm   5 cm

c   15 cm   20 cm

2 Lisa is making decorative book ends for presents.
She buys a square of wood and cuts it in half.
This table shows how much the wood costs:

Area of wood	Cost
Less than 200 cm²	50p
200 cm² to 600 cm²	£1
Over 600 cm²	£1.75

How much will Lisa pay for each of these?

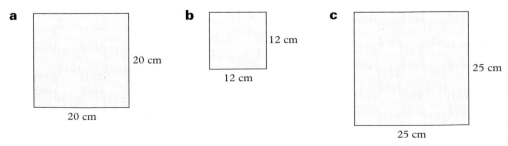

a   20 cm   20 cm

b   12 cm   12 cm

c   25 cm   25 cm

3   Hannah paints logos on sweatshirts.
Her charges depend on the size of the logo.

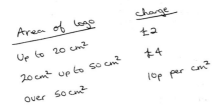

Area of logo	charge
Up to 20 cm²	£2
20 cm² up to 50 cm²	£4
over 50 cm²	10p per cm²

How much would each of these logos cost?

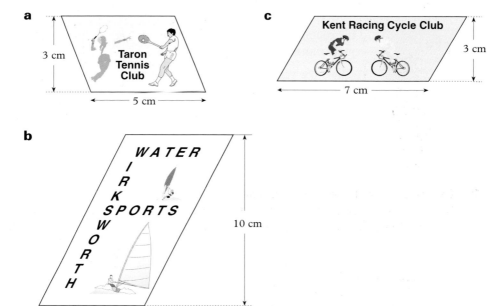

a
3 cm
Taron Tennis Club
5 cm

c
Kent Racing Cycle Club
3 cm
7 cm

b
WATER
I R K SPORTS W O R T H
10 cm
6 cm

3 Areas of triangles

The total area of the sails of a yacht is important. The bigger the area, the faster the yacht will sail. If the area is too big, the yacht could break under the strain.

| **Area of a triangle** |

This rectangle is divided into two equal triangles.

Area of the rectangle = base × height

So the area of the triangle is half the area of the rectangle.

Area of a triangle $= \dfrac{\text{base} \times \text{height}}{2}$

Example

Find the area of this triangle.

Area of a triangle $= \dfrac{\text{base} \times \text{height}}{2}$

$\qquad\qquad\quad = \dfrac{6 \times 4}{2}$

$\qquad\qquad\quad = 12 \text{ cm}^2$

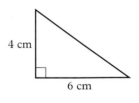

Exercise 1:8

Find the areas of these triangles.

1

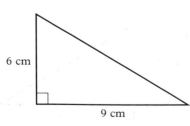

2

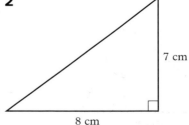

3

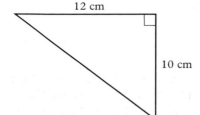

12 cm

10 cm

4

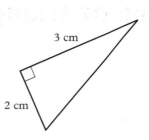

3 cm

2 cm

This parallelogram is divided into two equal triangles.

Area of a parallelogram = base × height

So the area of the triangle is half the area of the parallelogram.

Area of a triangle = $\dfrac{\text{base} \times \text{height}}{2}$

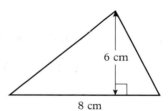

height

base

Example   Find the area of this triangle.

Area of a triangle = $\dfrac{\text{base} \times \text{height}}{2}$

$= \dfrac{8 \times 6}{2}$

$= 24 \text{ cm}^2$

6 cm

8 cm

Exercise 1:9

Find the areas of these triangles.

1

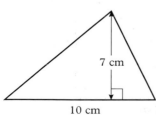

7 cm

10 cm

2

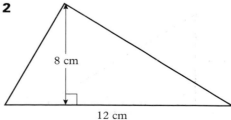

8 cm

12 cm

3

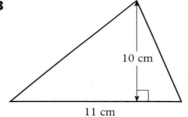

10 cm

11 cm

4

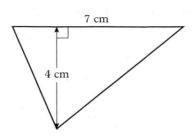

7 cm

4 cm

5 **a** Find the area of each of these two sails.
b Which sail is larger?

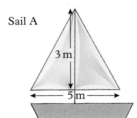

Sail A

3 m

5 m

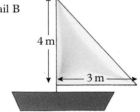

Sail B

4 m

3 m

6 These triangles are drawn on axes.
Find the area of each triangle.

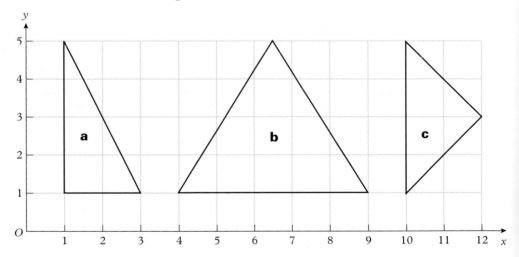

7 This is a plan of a field.
QS is a footpath going straight across the field.
It is 60 m long.
   a Find the area of triangle PQS.
   b Find the area of triangle QRS.
   c Find the total area of the field.

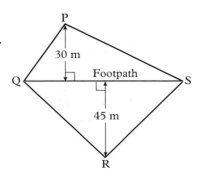

P

30 m

Footpath

Q

S

45 m

R

1   Write down the area of each shape.

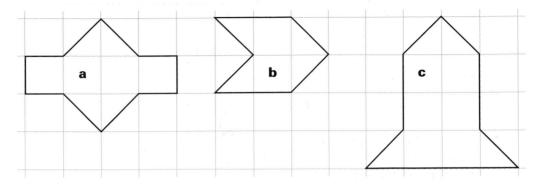

2   The area of the square is 9 cm². The area of the triangle is 5 cm².

Find the area of each shape.

a

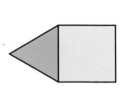

b

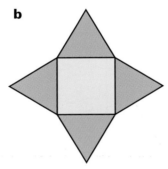

c

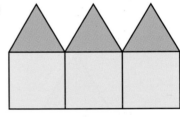

3   Find the areas of these rectangles.

a

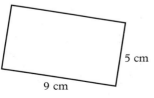

9 cm
5 cm

b

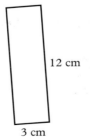

12 cm
3 cm

c

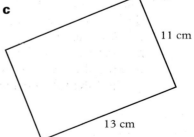

11 cm
13 cm

19

4 This yacht has two sails.
- **a** Find the area of each sail.
- **b** Find the total area of the sails.

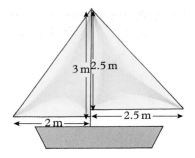

3 m 2.5 m

2 m

2.5 m

5 Find the areas of these parallelograms.

a

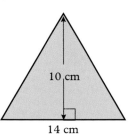

10 cm

9 cm

b

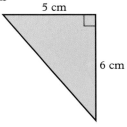

13 cm

5 cm

c

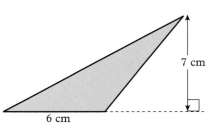

9 cm

12 cm

10 cm

6 Find the areas of these triangles.

a

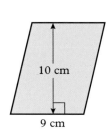

10 cm

14 cm

b

5 cm

6 cm

c

7 cm

6 cm

7 **a** Find the area of each of these logos.
   b Which has the smallest area?

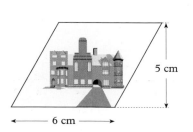

5 cm

6 cm

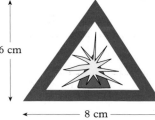

6 cm

8 cm

Cubley Cricket Club

7 cm

8 cm

8 These shapes are drawn on axes.
Find the area of each shape.

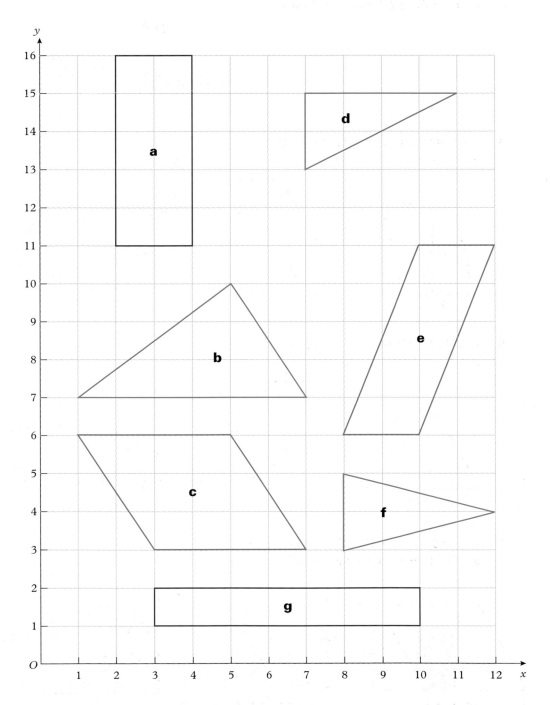

1   These three shapes have the same area.

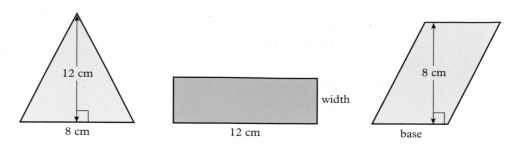

a   Find the area of the triangle.
b   Find the width of the rectangle.
c   Find the base of the parallelogram.

2   **a**   Find the area of
rectangle P.
b   Find the area of
rectangle Q.
c   Find the total area
of the shape.

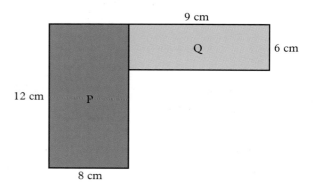

3   Find the area of each shape.

a

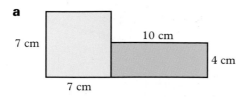

b

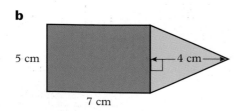

- **Area**   **Area** is measured using squares.

  1 cm² is the area of a
  square 1 cm by 1 cm.

  This shape has 5 squares.
  The area of this shape is 5 cm².

- **Area of a
  rectangle**

  Area of a rectangle = *l*ength × *w*idth
  $$= l \times w$$

  Example   Calculate the area of this
  rectangle.

  Area = length × width
  $$= 7 \times 3$$
  $$= 21 \text{ cm}^2$$

  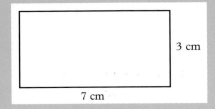
  3 cm

  7 cm

- **Area of a
  parallelogram**

  Area of a parallelogram = Area of a rectangle
  $$= \text{base} \times \text{height}$$

  Example   Find the area of this
  parallelogram.

  Area = base × height
  $$= 7 \times 4$$
  $$= 28 \text{ cm}^2$$

  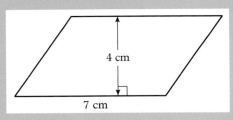
  4 cm

  7 cm

- **Area of a
  triangle**

  Area of a triangle $= \dfrac{\text{base} \times \text{height}}{2}$

  Example   Find the area of this triangle.

  Area of a triangle $= \dfrac{\text{base} \times \text{height}}{2}$

  $$= \frac{6 \times 4}{2}$$

  $$= 12 \text{ cm}^2$$

  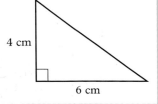
  4 cm

  6 cm

1 Write down the area of each shape.

a
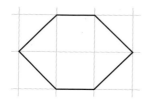

b
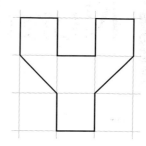

2 Find the area of each of these shapes.

a
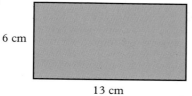
6 cm
13 cm

c
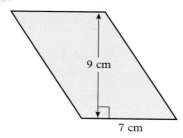
9 cm
7 cm

b
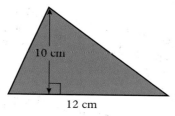
10 cm
12 cm

d
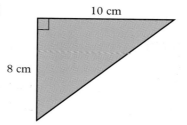
10 cm
8 cm

3 a Find the area of each sail.
b Find the area of the hull.

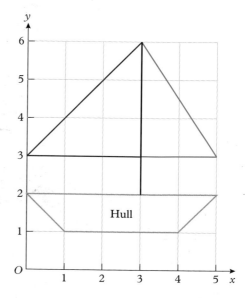

2 Number

Before decimalisation took place in the UK in 1971 one pound sterling was made up of 20 shillings, with each shilling made up of 12 pence. How many pence were there in a pound?

The new system changed the number of pence in one pound to 100.

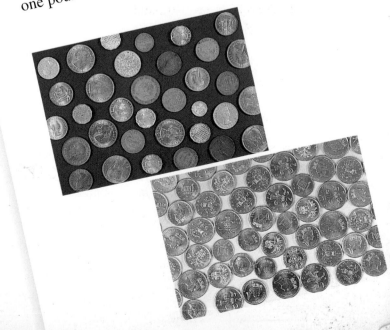

1 Addition and multiplication

Tracy takes four oranges to a checkout.
She gets a checkout slip like this:
The oranges cost 16 p each.
16 p is £0.16
On the slip it is *0.16*

ORANGES	0.16
ORANGES	0.16
ORANGES	0.16
ORANGES	0.16
TOTAL 0.64	

Peter buys four oranges at a different checkout.
Peter's checkout slip looks like this:

ORANGES	
4 @ 0.16	0.64
TOTAL 0.64	

Exercise 2:1

1 This is Carl's checkout slip:
The *TOTAL* has got torn off.

   a Find Carl's *TOTAL* by adding.

COLA	0.28
COLA	0.28
COLA	0.28
TOTAL	

   b Here is a new checkout slip for Carl.
      Fill it in.

COLA	
3 @ 0.28	...
TOTAL	...

For each of the checkout slips below:
a Find the *TOTAL* by adding.
b Make a different checkout slip using multiplying.
c Check the *TOTAL* by multiplying.

2

CAT FOOD	0.39
CAT FOOD	0.39
CAT FOOD	0.39
CAT FOOD	0.39
TOTAL	...

4

COLA	0.28
COLA	0.28
COLA	0.28
COLA	0.28
COLA	0.28
TOTAL	...

3

PIZZA	2.75
PIZZA	2.75
PIZZA	2.75
TOTAL	...

5

CHICKEN	3.26
CHICKEN	3.26
TOTAL	...

It is often quicker to work out one multiplication than to do several additions.

Example

 + +

a Write this addition as a multiplication.
b Work out the answer.

a 3 × 4   b 12 biscuits

Exercise 2:2

a Write these additions as multiplications.
b Work out the answers.

1 How many cakes are there altogether?

 + + +

2 How many sweets are there altogether?

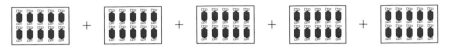

3 1500 + 1500 + 1500 + 1500

5 8 + 8 + 8 + 8 + 8 + 8

4 203 + 203 + 203

6 17 + 17 + 17 + 17 + 17

Example

Find the number of mince pies in three boxes of 6 pies.

a Write the working as an addition.
b Write the working as a multiplication.
c Work out the answer.

a 6 + 6 + 6   **b** 3 × 6   **c** 18

---

7 Find the number of eggs in four boxes.
Each box has 12 eggs in it.
a Write the working as an addition.
b Write the working as a multiplication.
c Work out the answer.

8 Find the number of seats in four coaches.
Each coach has 53 seats.
a Write the working as an addition.
b Write the working as a multiplication.
c Work out the answer.

9 Find the number of sheets of paper in three
packets.
There are 500 sheets in a packet.
a Write the working as an addition.
b Write the working as a multiplication.
c Work out the answer.

2 Subtraction and division

Karen and Paul are filling egg boxes.
Each box takes 6 eggs.
Karen and Paul have 30 eggs each.

Karen wants to know how many
boxes she can fill with her eggs.
She works the answer out like this:

first box	30 − 6 = 24 left
second box	24 − 6 = 18 left
third box	18 − 6 = 12 left
fourth box	12 − 6 = 6 left
fifth box	6 − 6 = 0 left

Karen can fill five boxes with her eggs.

Paul also works out how many boxes
he can fill.
He does it like this:

30 ÷ 6 = 5

Paul gets the same answer as Karen.

Exercise 2:3

1 Sharon is filling trays with apples.
Each tray holds 4 apples.
Sharon starts with 28 apples.
She works out how many trays she can fill like this:

28 − 4 = 24   16 − 4 = 12   4 − 4 = 0
24 − 4 = 20   12 − 4 = 8
20 − 4 = 16   8 − 4 = 4

a How many times does Sharon take away 4?
b How many trays does Sharon fill with 4 apples?
c Check your answer using this division:
28 ÷ 4 = ...

2   Jason is putting oranges into nets.
Each net holds 5 oranges.
Jason starts with 30 oranges.
He works out how many nets he can fill like this:

$30 - 5 = 25$ $20 - 5 = 15$ $10 - 5 = 5$
$25 - 5 = 20$ $15 - 5 = 10$ $5 - 5 = 0$

a   How many times does Jason take away 5?
b   How many nets does Jason fill with 5 oranges?
c   Check your answer using this division:
$30 \div 5 = \ldots$

3   Allison is putting peaches into boxes.
Each box holds 7 peaches.
Allison wants to know how many boxes she can fill.
She starts with 28 peaches.
a   Copy these subtractions.
Take away 7 until you get 0.

$28 - 7 = 21$
$21 - 7 = \ldots$

b   How many times do you take away 7?
c   How many boxes does Allison fill with peaches?
d   Check your answer using this division:
$28 \div 7 = \ldots$

4   Andrew is putting bread rolls into bags of 8.
He starts with 40 rolls.
a   Copy these subtractions.
Continue taking away 8 until you get 0.

$40 - 8 = 32$
$32 - \ldots = \ldots$

b   How many bags does Andrew fill?
c   Check your answer using this division:
$40 \div 8 = \ldots$

John is putting 34 kiwi fruit into boxes of 8.

John uses his calculator to do this division:
$$34 \div 8 = 4.25$$
John can fill **4** boxes with kiwi fruit.

John works out the number of kiwi fruit
left over like this:
$$34 - 8 - 8 - 8 - 8 = 2$$
John takes away **8** *four* times.
He is left with 2.
2 kiwi fruit are left over.

Exercise 2:4

1 Jane has 23 pears.
She is putting them into boxes of 4.
Jane does this division:
$$23 \div 4 = 5.75$$

   a How many boxes does Jane fill with 4 pears?
   b How many times does Jane take away 4
      to find the remainder?
   c Copy and complete this subtraction:
$$23 - 4 - 4 - \ldots$$
   d How many pears are left over?

2 Sally has 26 bottles of washing-up liquid.
She puts them into boxes of 6.
Sally does this division:
$$26 \div 6 = 4.3333\ldots$$

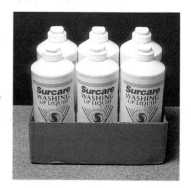

   a How many boxes can Sally fill with 6 bottles?
   b How many times does Sally take away 6 to
      find the remainder?
   c Copy and complete this subtraction:
$$26 - 6 - 6 - \ldots$$
   d How many bottles are left over?

3 Jason is cooking sausages for 7 scouts.
Jason starts with 30 sausages.
Jason does this division:

   30 ÷ 7

Copy Jason's working.
Fill it in.

   30 ÷ 7 = ...
   Each scout gets ... whole sausages.

   30 − 7 − 7 − ... = ...
   There are ... sausages left.

4 Melissa is packing onions into nets.
A net holds 8 onions.
Keith does this division:

   36 ÷ 8

Copy Melissa's working.
Fill it in.

   36 ÷ 8 = ...
   There are ... nets of onions.

   36 − 8 − 8 − ... = ...
   There are ... onions left over.

5 Navdeep is tying old magazines in bundles
of 12 for the school fair.
Navdeep starts with 40 magazines.
   a Use your calculator.
   Do a division.
   Find the number of bundles Navdeep makes.
   b Do a repeated subtraction.
   Find the number of magazines left over.

Robert packs pizzas into packs of 6.

He starts with 535 pizzas.
Robert wants to know how many packs he will get and how many will be left over.

Here is Robert's working:
   $535 \div 6 = \mathbf{89}.1666 \ldots$
Robert will get **89** packs of pizzas.

Robert needs to take away 6 on his calculator 89 times to find the number left over.
Robert knows this would take too long!

Robert works out the number left over like this:
   89 packs hold 89×6 pizzas = 534
   Number left over: $535 - 534 = \mathbf{1}$
Robert has **1** pizza left over.

Exercise 2:5

For questions **1** to **4** work out:

a the number of whole packs,
b the number left over.

1 A machine counts 3500 tea bags into packs of 40.
Copy and complete:
   $3500 \div 40 = 87.5$
There are ... packs of 40 tea bags.
... packs hold ... $\times$ 40 = ... tea bags
Number left over: 3500 − ... = ...

2 A machine counts 1250 biros into packs of 12.

3 A machine counts 5000 tins of beans into packs of 24.

4 A machine counts 1845 colas into packs of 6.

5 Find the remainder in each of these divisions:
   a $950 \div 17$ **b** $2054 \div 7$ **c** $8000 \div 23$ **d** $172 \div 3$

Broken calculators

Exercise 2:6

Most of 9P's calculators have a key missing.
Do these on your calculator
without using the missing keys!
Write down your answers.

1 Keith's calculator has **5** missing.
He needs to work out 17×5.

2 Anna's calculator has no **÷**.
Anna needs to work out $56 \div 14$.

3 David's calculator has **×** missing.
He needs to work out 34×6.

4 Vicky's calculator has no **+**.
She needs to work out $18 + 18 + 18 + 18$.

5 Helen's calculator has a broken **7**.
She needs to work out 49×7.

6 Dale's calculator has no **×**.
He needs to work out 45×8.

● **7** Jo's calculator has the **5** missing.
She needs to work out 15×7.

● **8** Andy's calculator has no **4**.
He needs to work out 24×6.

● **9** Harry's calculator has a broken **÷** and a broken **4**.
He needs to work out $54 \div 6$.

● **10** Barbara's calculator has both the **÷** and the **8** missing.
She needs to work out $48 \div 16$.

Make up some more of your own.

3 Working without a calculator

Addition and subtraction

Examples

Work these out in your head.
Write down the answers.

$$1 + 4 = 5$$
$$11 + 4 = 15$$
$$21 + 14 = 35 \ (2 + 1 = 3 \text{ and } 1 + 4 = 5)$$

Exercise 2:7

Work these out in your head.
Write down the answers.

1 **a** $2 + 3 =$ **b** $3 + 4 =$ **c** $3 + 5 =$ • **d** $3 + 21 =$
     $12 + 3 =$ $13 + 4 =$ $15 + 3 =$ $14 + 32 =$
     $2 + 13 =$ $3 + 24 =$ $23 + 15 =$ $25 + 13 =$
     $12 + 13 =$ $23 + 14 =$ $13 + 35 =$ $24 + 11 =$

2 **a** $4 + 5 =$ **b** $5 + 2 =$ **c** $4 + 2 =$ • **d** $12 + 35 =$
     $4 + 15 =$ $15 + 12 =$ $12 + 14 =$ $14 + 25 =$
     $14 + 5 =$ $25 + 22 =$ $24 + 22 =$ $42 + 13 =$
     $14 + 15 =$ $15 + 32 =$ $32 + 34 =$ $51 + 12 =$

'Shopkeepers' addition'

You can do subtractions by 'counting on' in your head.
Shopkeepers do this with money – when you give shopkeepers money,
they can work out your change by 'counting on' from the price of the
goods to the amount you gave them.

Examples   **1**   Find the change when 15 p is paid using a 20 p coin.

Count on **five units** from 15 p to get to 20 p   **16, 17, 18, 19, 20**
The change is 5 p

2   Find the change when 26 p is paid using a 50 p coin.

Count on **four units** from 26 p to get to 30 p   **27, 28, 29, 30**
Count on two tens from 30 p to get to 50 p   40, 50
The change is 24 p

---

3   Work these out in your head.
Write down the answers.

a 13 − 5 =	**b** 25 − 19 =	**c** 34 − 26	• **d** 35 − 18
15 − 6 =	21 − 15 =	31 − 23	32 − 13
17 − 9 =	23 − 18 =	37 − 24	36 − 11
19 − 12 =	26 − 17 =	36 − 29	33 − 14

4   Write down what you get when you take these numbers away from 10.
    a 2   **b** 7   **c** 6   **d** 3   **e** 8

5   Write down what you get when you take these numbers away from 20.
    a 17   **b** 16   **c** 4   **d** 8   **e** 5

6   Write down what you get when you take these numbers away from 30.
    a 23   **b** 19   **c** 12   **d** 5   **e** 3

7   3 + 9 = 12 and 9 + 3 = 12
3 and 9 are a pair of numbers that add up to 12.
Write down all the pairs of numbers that add up to 12.

8   Write down all the pairs of numbers that add up to
    a 14   **b** 16   **c** 11   **d** 18   **e** 21

Multiplication

You can multiply without using a calculator.

Example

Set out 41 × 5 as   41   first do   41   then do   41
             × 5   5 × 1   × 5   5 × 4   × 5
                                           5   205

Exercise 2:8

Work these out.

1   23 × 2   **2**   51 × 3   **3**   62 × 4   **4**   71 × 5

Sometimes you need to carry.

Example   25   →   25
                 × 3   × 3
                  5₁   75₁

$3 × 2 = 6$

Then add the **1** to give 7

Work these out.

5   14 × 5   **6**   24 × 4   **7**   26 × 3   **8**   47 × 2

Set these out in the same way.

9   34 × 5   **11**   27 × 7   **13**   63 × 9   ● **15**   403 × 7

10   48 × 4   **12**   36 × 8   **14**   81 × 6   ● **16**   246 × 9

When you want to multiply two quite large numbers you have to do it in stages.
There is another method in Help Yourself.
You only have to know one method.

Example 163×42

First do 163×2 then do 163×40 Now add the two answers together.

$$\begin{array}{r} 163 \\ \times\ \ 2 \\ \hline 326 \\ \tiny 1 \end{array} \qquad \begin{array}{r} 163 \\ \times\ \ 4 \\ \hline 652 \\ \tiny 2\ 1 \end{array} \qquad \begin{array}{r} 326 \\ +\ 6520 \\ \hline 6846 \end{array}$$

$$652 \times 10 = 6520$$

Usually the working out looks like this:

$$\begin{array}{r} 163 \\ \times\ \ 42 \\ \hline 326 \\ \tiny 1 \\ 6520 \\ \tiny 2\ 1 \\ \hline 6846 \end{array}$$

Here is another example.
364×32

$$\begin{array}{r} 364 \\ \times\ \ 32 \\ \hline 728 \\ \tiny 1 \\ 10\,920 \\ \tiny 1\ \ 1 \\ \hline 11\,648 \\ \tiny 1 \end{array}$$

$\leftarrow (364 \times 2)$
$\leftarrow (364 \times 30)$

Exercise 2:9

Use the method you prefer to work these out.

1 142×34	**4** 68×82	**7** 67×29	**10** 34×27
2 243×52	**5** 308×62	**8** 512×38	**11** 360×56
3 312×67	**6** 416×91	**9** 406×84	**12** 129×36

Division

You can divide without using a calculator.

Example $86 \div 2$ $2\overline{)86}$

First work out $8 \div 2 = 4$. $\dfrac{4}{2\overline{)86}}$
Put the 4 above the 8:

Now work out $6 \div 2 = 3$. $\dfrac{43}{2\overline{)86}}$
Put the 3 above the 6.

Exercise 2:10

Work these out.

1 $3\overline{)69}$ **2** $2\overline{)64}$ **3** $4\overline{)48}$ **4** $2\overline{)82}$

Sometimes you have to 'carry'. This happens when a number does not divide exactly.

Example $96 \div 4$ $4\overline{)96}$

First do the $9 \div 4$. This is 2 with 1 left over.
Put the 2 over the 9 and carry the 1 like this: $\dfrac{2}{4\overline{)9^16}}$

Now do 16 divided by 4. This is 4.
Put the 4 above the 6 like this: $\dfrac{2\ 4}{4\overline{)9^16}}$

So $96 \div 4 = 24$

Work these out.

5 $5\overline{)65}$ **7** $318 \div 2$ **9** $78 \div 2$ **11** $534 \div 6$

6 $3\overline{)57}$ **8** $132 \div 3$ **10** $312 \div 4$ ● **12** $535 \div 5$

Sometimes there is a remainder left at the end.

Example $138 \div 4$

$\dfrac{3\ 4}{4\overline{)13^18}}$ remainder 2   ← ($18 \div 4 = 4$ remainder 2)

These divisions have remainders.
Work out the divisions.

13 $37 \div 2$ **15** $146 \div 3$ **17** $321 \div 2$ **19** $507 \div 2$

14 $136 \div 5$ **16** $154 \div 4$ **18** $475 \div 4$ **20** $829 \div 3$

Sometimes you need to do long division.

Example $432 \div 12$

12 will not go into 4 so first do $42 \div 12$ $12\overline{)420}$

You need to find out how many times 12 goes into 42. $12 \times 2 = 24$
                                                       $12 \times 3 = 36$ ←
                                                       $12 \times 4 = 48$

12 will go in **3** times. $\dfrac{3}{12\overline{)420}}$
Put the **3** above the 2. 36
$12 \times 3 = 36$
Put **36** under the 42.

                                              $\dfrac{3}{12\overline{)420}}$
Now subtract 36 from 42. 36
                                               $\overline{6}$

The **6** is the 'carry'. $\dfrac{3}{12\overline{)420}}$
You don't put it with the 0. $36\downarrow$
Instead you bring down $\overline{60}$
the **0** to the **6**.

Now do $60 \div 12$ $12 \times 4 = 48$
                                               $12 \times 5 = 60$ ←

12 will go **5** times exactly. $\dfrac{35}{12\overline{)420}}$
Put the **5** in after the 3. 36
                                               $\overline{60}$

$12 \times 5 = 60$ $\dfrac{35}{12\overline{)420}}$
Put the answer under the 60. 36
When you subtract this time there is no remainder. $\overline{60}$
                                                              60
This is what the working looks like $\overline{60}$
when you have finished!

So $420 \div 12 = 35$

Exercise 2:11

Work these out by long division.
They should all work out exactly.

1 $312 \div 12$ **3** $714 \div 21$ **5** $481 \div 13$

2 $510 \div 15$ **4** $650 \div 25$ **6** $1472 \div 32$

7 $810 \div 18$ **9** $672 \div 14$ **11** $1428 \div 42$

8 $1768 \div 34$ **10** $851 \div 23$ ● **12** $2337 \div 19$

Exercise 2:12

1   Alice sells bird tables for £24 each.
How much does Alice get if she sells 35 bird tables?

2   Alice also sells bird baths.
She gets £414 for 18 bird baths.
How much does Alice charge for one bird bath?

Diagonals

1   This small square of 4 numbers
comes from the larger square.

    a   What do you notice about
$13 + 24$ and $14 + 23$?

13	14
23	24

    b   Try other four-number squares.
Is there a rule?

    c   Try a larger square.
Does this square
follow your rule?

42	43	44
52	53	54
62	63	64

1	2	3	4	5	6	7	8	9	10
11	12	13	14	15	16	17	18	19	20
21	22	23	24	25	26	27	28	29	30
31	32	33	34	35	36	37	38	39	40
41	42	43	44	45	46	47	48	49	50
51	52	53	54	55	56	57	58	59	60
61	62	63	64	65	66	67	68	69	70
71	72	73	74	75	76	77	78	79	80
81	82	83	84	85	86	87	88	89	90
91	92	93	94	95	96	97	98	99	100

2   What do you notice about
13×24 and 14×23?
Try some other squares.

3   Do the same rules work for rectangles?

1   For each of the checkout slips below:
  a Find the *TOTAL* by adding.
  b Make a new checkout slip using multiplying.
  c Check the *TOTAL* by multiplying.

MEAT PIE	1.26
MEAT PIE	1.26
MEAT PIE	1.26
MEAT PIE	1.26
TOTAL	...

ORANGE	0.18
ORANGE	0.18
ORANGE	0.18
ORANGE	0.18
ORANGE	0.18
TOTAL	...

CRISPS	0.24
CRISPS	0.24
CRISPS	0.24
TOTAL	...

2   Write these additions as multiplications.
  Work out the answers.

  a   [image] + [image] + [image] + [image] + [image]

  b   156 + 156 + 156 + 156 + 156 + 156 + 156 + 156
  c   2000 + 2000 + 2000

3   Find the number of felt tips in five boxes
  of 10 felt tips.
  a Write the working as an addition.
  b Write the working as a multiplication.
  c Work out the answer.

4   Find the number of colas in four packs
  of 15 cans.
  a Write the working as an addition.
  b Write the working as a multiplication.
  c Work out the answer.

5 Michelle is putting pitta bread into packs of six.
She starts with 45 pittas.
Michelle wants to know how many packs she can fill and how many pittas she will have left over.

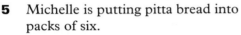

  a Use your calculator to do a division.
  Find how many packs of pittas Michelle makes.
  b Use your calculator to find how many Michelle has left over.

6 For each of these work out the number of packs and the number left over.
  a 2750 chocolates in boxes of 36
  b 1750 sausages in packs of eight

7 Write down all the pairs of numbers that add up to:
  a 13   **b** 19   **c** 15   **d** 17

8 Write down what you get when you take these numbers away from 30.
  a 27   **b** 19   **c** 16   **d** 4   **e** 8

9 Work these out.
  a $17 - 8 =$ **c** $23 - 14 =$ **e** $35 - 26 =$ **g** $27 - 18 =$
  b $32 - 27 =$ **d** $25 - 18 =$ **f** $42 - 33 =$ **h** $44 - 37 =$

10 Work these out.
  a 46×5 **c** 89×2 **e** 57×16 **g** 253×37
  b 52×7 **d** 67×4 **f** 145×24 **h** 93×45

11 Work these out.
They divide exactly.
  a $168 \div 7$ **b** $315 \div 5$ **c** $336 \div 6$ **d** $824 \div 4$

12 Work these out.
They have remainders.
  a $124 \div 5$ **b** $146 \div 3$ **c** $527 \div 4$ **d** $179 \div 8$

13 Work these out.
They divide exactly.
  a $345 \div 15$ **b** $1034 \div 22$ **c** $2015 \div 31$ **d** $1368 \div 24$

1 Here is Harneet's checkout slip.
  a Find the *TOTAL* by adding.
  b Make a new checkout slip using multiplying.
  c Check the *TOTAL* by multiplying.

DOG FOOD	0.43
DOG FOOD	0.43
DOG FOOD	0.43
SAUSAGES	1.25
SAUSAGES	1.25
TOTAL	...

2 Mr Jones is putting the 177 pupils in Year 9 into classes.
He wants 29 pupils in each class.
  a Use division to find the number of classes in Year 9.
  b Use repeated subtraction to find the number of pupils left over.
  c How many classes in Year 9 will have 29 pupils?
  How many classes will need to have 30 pupils?

3 Sherene's machine is putting yogurts in packs of 8.
Sherene has 2550 yogurts.
She works out how many packs of yogurts she will get and how many
will be left over.
Use your calculator to work out Sherene's answers.

4 Mark makes rocking horses to sell.
He sells the rocking horses for £134 each.
  a Mark sells 26 rocking horses.
  How much does he get for the 26 horses?
  b Mark has a box of 250 brass studs.
  He uses 14 brass studs on each horse.
  How many complete horses can
  he make using the 250 brass studs?

5 Do these by long division.
There are remainders in these questions.
  a 158 ÷ 12   **b** 1010 ÷ 19   **c** 1500 ÷ 23   **d** 3112 ÷ 42

- It is often quicker to work out one multiplication than to do several additions.

  Example

  a Write this addition as a multiplication.
  b Work out the answer.

  a 3×4 **b** 12 biscuits

- John is putting 60 kiwi fruit into boxes of 8.

  John uses his calculator to do this division:
  $60 \div 8 = 7.5$
  John can fill **7** boxes with kiwi fruit.

  John works out the number of kiwi fruit left over like this:
  $60 - 8 - 8 - 8 - 8 - 8 - 8 - 8 = 4$
  John takes away **8** *seven* times. 4 kiwi fruit are left over.

- Robert wants to know how many packs of 6 he will get from 535 pizzas and how many will be left over.

  Here is Robert's working: Robert works out the number left over like this:
  $535 \div 6 = \mathbf{89}.1666 \ldots$ 89 packs hold 89×6 pizzas $= 534$
  Robert will get **89** packs   Number left over: $535 - 534 = \mathbf{1}$
  of pizzas.   Robert has **1** pizza left over.

- You need to be able to work without a calculator.

  Examples   Work these out in your head.
  Write down the answers.

  $1 + 4 = 5$ $11 + 4 = 15$ $21 + 14 = 35$

- Multiplying or dividing two large numbers is easier in stages.

  Examples   **1** 364×32 **2** $420 \div 12$

$$
\begin{array}{r}
364 \\
\times \quad 32 \\
\hline
728 \\
10\,920 \\
\hline
11\,648 \\
\hline
\end{array}
\begin{array}{l}
\leftarrow (364 \times 2) \\
\leftarrow (364 \times 30)
\end{array}
$$

$$
\begin{array}{r}
35 \\
12\overline{)420} \\
36\downarrow \\
\hline
60 \\
60 \\
\hline
\end{array}
$$

1 **a** Write this addition as a multiplication.
    $14 + 14 + 14 + 14 + 14 + 14 + 14$
b Work out the answer.

2 Martin's machine is shrink wrapping
cans of cola into packs of 4.
Martin wants to know how many packs
he will get from 35 cans of cola and
how many will be left over.
Here is Martin's working:
    $35 \div 4 = 8.75$
a How many packs of 4 will Martin get?
b How many times does Martin need to
    take away 4 to get the remainder?
c How many cans are left over?

3 Kamal has 1525 bread rolls to put in packs of 6.
a How many packs will she get?
b How many rolls will be left over?

4 Do these in your head.
Write down the answers.
a $24 - 17$ **b** $33 - 24$ **c** $28 - 19$ **d** $18 - 9$

5 Take these numbers away from 40.
a 33   **b** 27   **c** 12   **d** 19   **e** 6

6 Work out these multiplications.
a 53×7 **b** 264×23

7 Work out these divisions.
They do not have remainders.
a $335 \div 5$ **b** $618 \div 3$ **c** $1242 \div 23$

8 Work out this division.
It does have a remainder.
    $86 \div 3$

3 Symmetry

CORE

1 Line symmetry
2 Rotational symmetry
3 Symmetry in 3-D

QUESTIONS

EXTENSION

SUMMARY

TEST YOURSELF

The photograph on the left shows Peter's face.

The photograph on the right shows what happens if the left half of Peter's face is reflected in a mirror line.

What are the differences between the two photographs?

1 Line symmetry

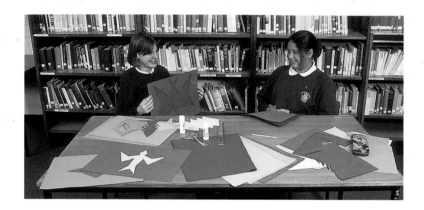

Exercise 3:1

1 **a** Fold a piece of paper in half.

b Draw a design on one side of your
paper like this:
It must start and finish on the fold.

c Cut along your design with scissors.

d Open out the paper.

e Put a mirror along the fold.
Look at the reflection in the mirror.
Look at the part of your design that
is behind the mirror.
They are the same.

2 Make some more designs.
Use the mirror to look at the reflection.

Image	The **image** is what we see in the mirror.
Line of symmetry	The fold is called the **line of symmetry** or mirror line.

A line of symmetry divides a shape into two equal parts.
If you fold the shape along this line, each part fits exactly on top of the other.

3 a Fold a piece of paper in half.
Fold it in half again.

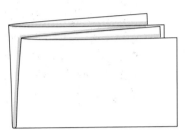

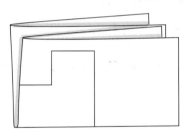

b Draw a design on your paper like this:
It must start and finish on the fold sides.

c Cut along your design with scissors.

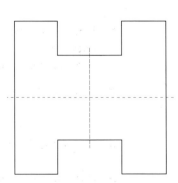

d Open out the paper.

e Put a mirror along one fold.
Look at the image in the mirror.
Do the same for the other fold.

f How many lines of symmetry does this design have?

4 Draw some more designs with two lines of symmetry.

W 5 You will need worksheet 3 : 1.

    a Cut out the equilateral triangle on the worksheet.
Use a mirror or fold the shape to find how many lines of symmetry the shape has.

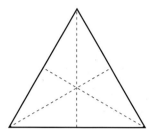

    Copy this table.
Fill in the first row.

Shape	Number of sides	Number of lines of symmetry
equilateral triangle		
square		
rectangle		
regular pentagon		
regular hexagon		
regular octagon		

    b Do the same for the other shapes.
Fill in the rest of the table.

    c The word **regular** means all the sides of the shape are the same length.
Which of these shapes is not regular?

    d Look at the regular shapes only.
What do you notice about the number of sides and the number of lines of symmetry?

W 6 You will need a piece of tracing paper and worksheet 3 : 2.

    a Make a tracing of the first shape.
Find out how many lines of symmetry the shape has by folding your tracing paper.
Write the answer on your worksheet.

    b Do the same for the rest of the shapes.

Exercise 3:2

    a Copy these shapes on to squared paper.
    b Mark on *all* the lines of symmetry.
        If there are no lines of symmetry, write 'none'.
    c Use tracing paper to check your answer.

1

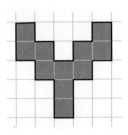

3

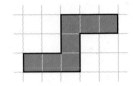

2

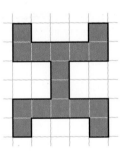

4

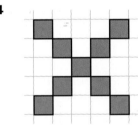

5

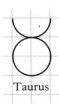

Taurus

7

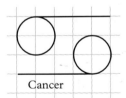

Cancer

9

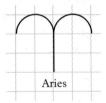

Aries

6

Pisces

8

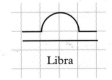

Libra

10

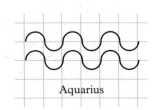

Aquarius

Example   You can only see half of this shape.
Use the line of symmetry to complete
the shape.

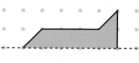

This is the complete shape:
Place a mirror along the line of
symmetry to check the answer.

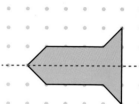

Exercise 3:3

1 Copy these shapes on to squared dotty paper.
Use the line of symmetry to complete these shapes.

a   **b**

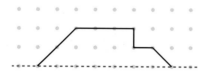   

2 Copy these shapes on to squared dotty paper.
Use the mirror line to complete these shapes.

a   **b**

3 Copy these patterns on to squared dotty paper.
Use the line of symmetry to complete the pattern.

a

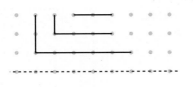

c

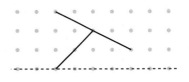

b

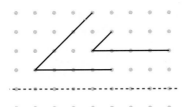

d

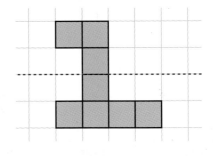

Example   Look at this pattern of shaded squares.
The dotted line is a line of symmetry.
Louise has to complete the pattern.

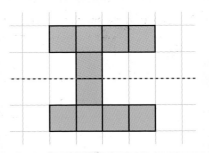

Louise has shaded in two more squares.
The pattern is now complete.

Exercise 3:4

1 Copy these shapes on to squared paper.
Shade in one more square to complete them.

a 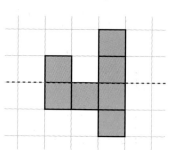 **b**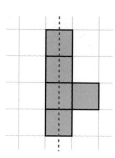

2 Copy these shapes on to squared paper.
Shade in two more squares to complete them.

a 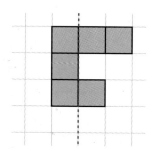 **b**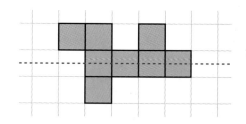

The mirror line is not always a line of the grid.

The reflection or image is still the same distance from the mirror line as the object.
Here the distance is diagonally across squares.

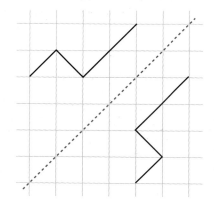

Copy these diagrams on to squared paper.
Draw their reflections in the mirror lines.
Use a mirror or tracing paper to help you.

3

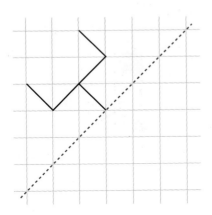

5

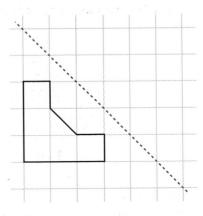

4

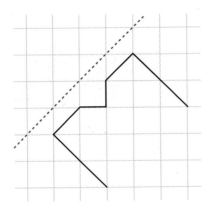

6

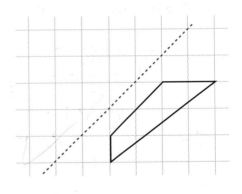

7 Copy these shapes on to squared paper.
Shade in two more squares to complete them.
Use a mirror or tracing paper to help you.

a

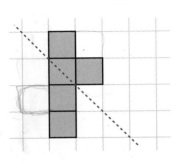

b

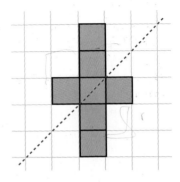

Colour can affect symmetry.

Example   This pattern has two lines of symmetry:

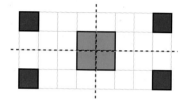

This pattern has only one line of symmetry:

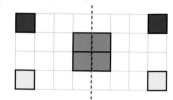

Exercise 3:5

1   These patterns have one line of symmetry. Copy and complete them.

a

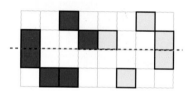

b

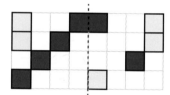

2   These patterns have two lines of symmetry. Copy and complete them.

a

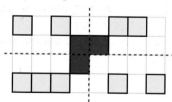

b

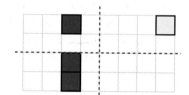

3   These patterns have four lines of symmetry. Copy and complete them.

a

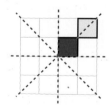

b

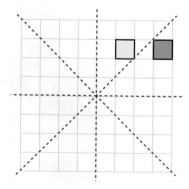

2 Rotational symmetry

 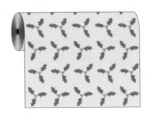 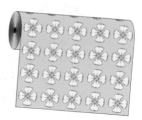

Wallpaper is designed with lots of symmetry.
As well as line symmetry, there is also rotational symmetry.

Rotate is another word for turn.
Rotational symmetry is to do with turning.

Angela traces this shape.
She puts a pencil point at the centre to hold the tracing paper in place.
She rotates the tracing paper through one complete turn.
The traced shape covers the original shape three times during the complete turn.
This shape has rotational symmetry of order 3.

Rotational symmetry	A shape has **rotational symmetry** if it fits on top of itself more than once as it makes a complete turn.
Order of rotational symmetry	The **order of rotational symmetry** is the number of times that the shape fits on top of itself. This must be two or more. Shapes that only fit on themselves once have no rotational symmetry.
Centre of rotation	The **centre of rotation** is the point at the centre of the shape.

Exercise 3:6

For each shape:
- **a** Trace the shape.
- **b** Use the tracing paper to see how many times the shape fits on top of itself in one complete turn.
- **c** Write down the order of rotational symmetry.
  If a shape has no rotational symmetry, write 'none'.

1

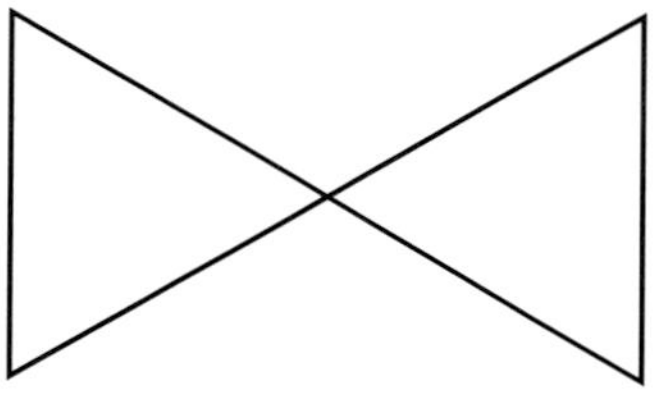

4

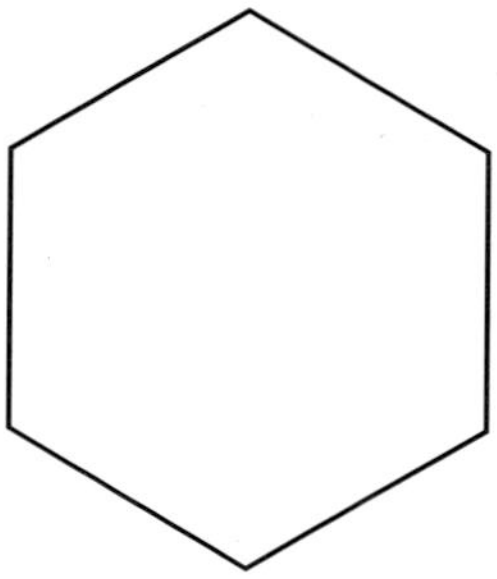

2

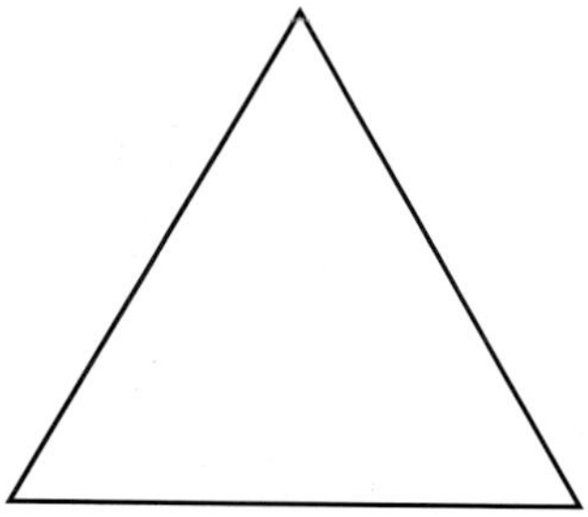

5

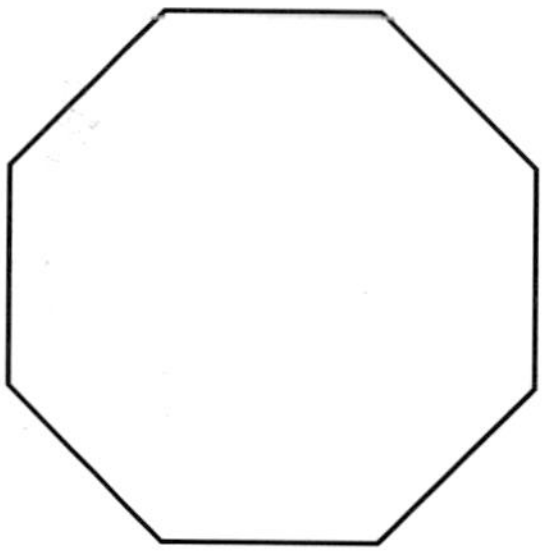

3

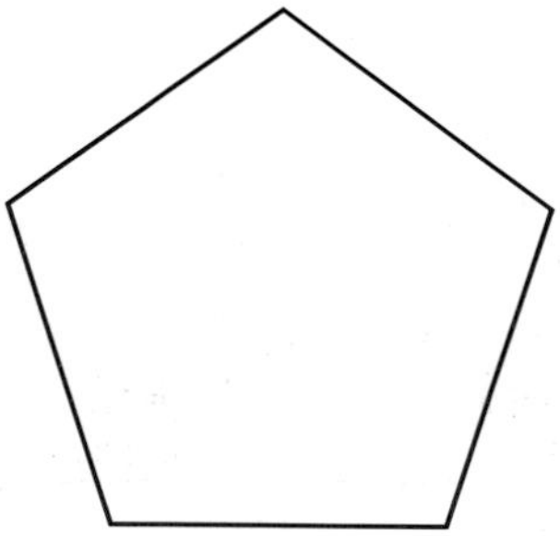

6

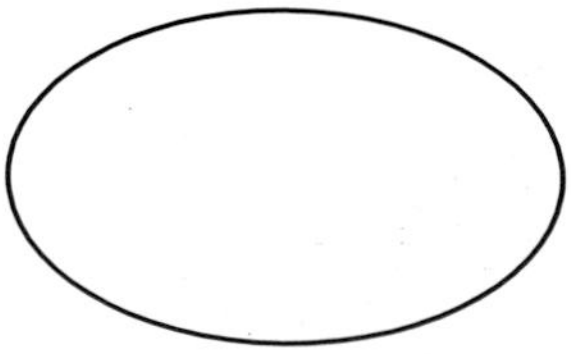

Example   This pattern looks the same after part of a turn.
In how many positions will the pattern look the same?
What is the order of rotational symmetry?

It will look the same in five different positions.
The order of rotational symmetry is 5.

Exercise 3:7

a In how many positions will each pattern look the same?
b Write down the order of rotational symmetry of each pattern.

1

3
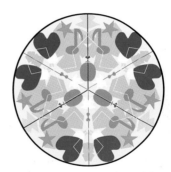

2

4
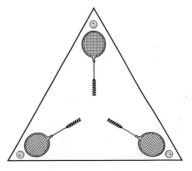

5 Design a logo that has rotational symmetry of order 4.
It can be for a sports club of your own choice.

3 Symmetry in 3-D

David has made some shapes with cubes.
All the shapes are symmetrical in 3-D.
He has drawn the shapes on isometric paper.

Isometric paper is printed in triangles.

It has a right way up.

right   wrong

Exercise 3:8

You will need isometric paper for this exercise.

1 Copy each of these diagrams on to isometric paper.

a   **b**   **c**

2 Draw each of these arrangements of cubes on to isometric paper.

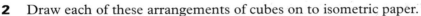

a  **b** 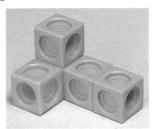 **c**

This shape is symmetrical about the mirror.

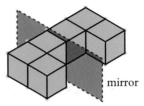

Example Complete this shape so that it is symmetrical on both sides of the mirror.

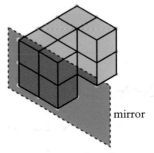

The completed shape looks like this:

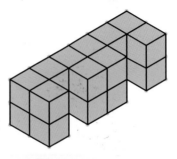

Exercise 3:9

Make each of these shapes with cubes.
Complete each shape so that it is symmetrical on both sides of the mirror.
Draw the complete shape on dotty isometric paper.

1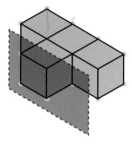

4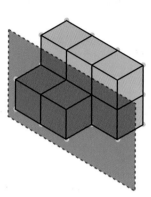

2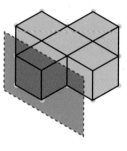

5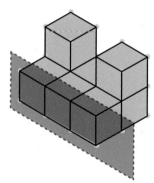

3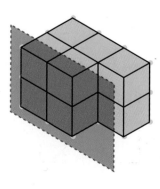

6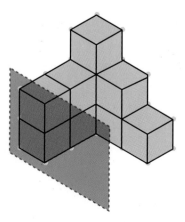

1 These are all symbols used in electronics.
Copy each symbol and mark on all the lines of symmetry.
If there are no lines of symmetry write 'none'.

a    **d** 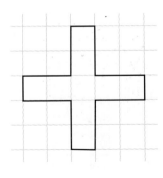   **g**

b   **e**   **h**

c   **f**   **i**

2 Copy these shapes on to squared paper.
Mark on all the lines of symmetry.

a 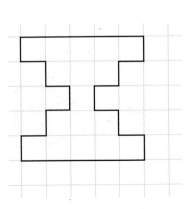   **c**

b 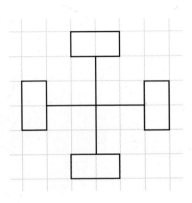   **d**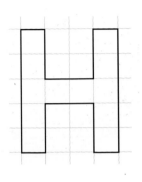

3 Copy these shapes on to squared paper.
Use the line of symmetry to complete each shape.

a

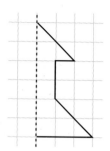

c

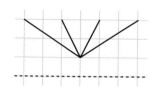

b

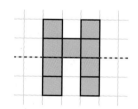

d

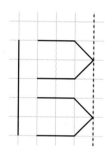

4 Copy these shapes on to squared paper.
Shade in one more square to complete them.

a

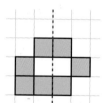

b

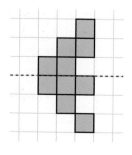

5 Copy these shapes on to squared paper.
Shade in two more squares to complete them.

a

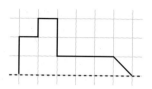

b

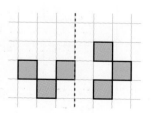

6 For each shape write down:
   a the number of times the shape fits on top of itself in one complete turn.
   b the order of rotational symmetry.

(1)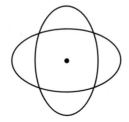

(3)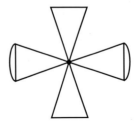

(2)

(4)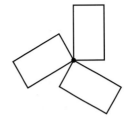

7 Copy these shapes on to squared paper.
   For each shape:
   a Mark on all the lines of symmetry.
   b Write down the order of rotational symmetry.

(1)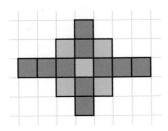

(3)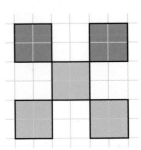

(2)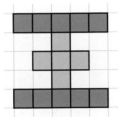

(4)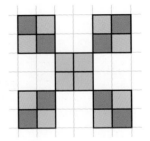

1 Copy these shapes on to squared paper.
   Use the line of symmetry to complete each shape.

   a

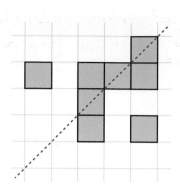

   c

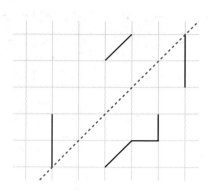

   b

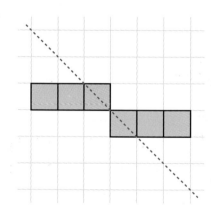

   d

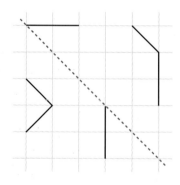

2 Draw each shape on dotty isometric paper so that it is symmetrical on
   both sides of the mirror.

   a

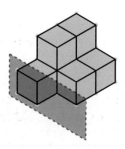

   b

- **Image**   The **image** is what we see in the mirror.

  Line of symmetry   The fold is called the **line of symmetry** or mirror line.

  A line of symmetry divides a shape into two equal parts. If you fold the shape along this line, each part fits exactly on top of the other.

- *Example*   You can only see half of this shape.
  Use the line of symmetry to complete the shape.

  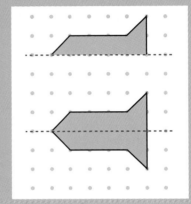

  This is the complete shape: Place a mirror along the line of symmetry to check the answer.

- *Example*   Colour can affect symmetry. This pattern has two lines of symmetry:

  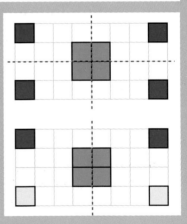

  This pattern has only one line of symmetry:

- **Rotational symmetry**   A shape has **rotational symmetry** if it fits on top of itself more than once as it makes a complete turn.

  Order of rotational symmetry   The **order of rotational symmetry** is the number of times that the shape fits on top of itself. This must be two or more. Shapes that only fit on themselves once have no rotational symmetry.

  Centre of rotation   The **centre of rotation** is the point at the centre of the shape.

1 Copy these shapes on to squared paper.
Mark on all the lines of symmetry.

a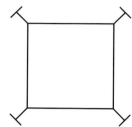

b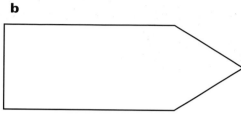

2 Copy these shapes on to squared paper.
Use the line of symmetry to complete the shape.

a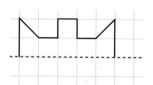

b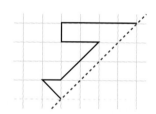

3 Copy these shapes on to squared paper.
Add squares to make the pattern complete.

a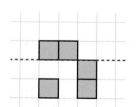

b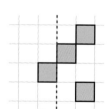

4 Write down the order of rotational symmetry of each of these shapes.

a

b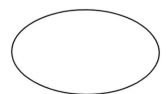

4 Statistics

QUESTIONS

EXTENSION

SUMMARY

TEST YOURSELF

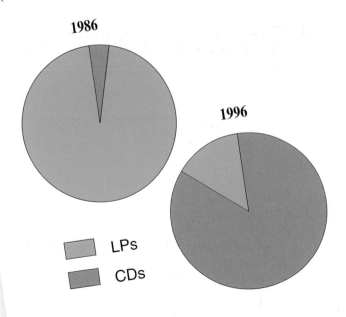

Relative share of retail sales in UK held by LPs and CDs (albums only) in 1986 and 1996.

1986

1996

LPs

CDs

Source: unpublished data.

1   Pictograms and bar-charts

Data is often shown in diagrams.
This makes it easier to understand.

Exercise 4:1

| **Pictogram** | A **pictogram** is a diagram which uses pictures. |
| **Key** | A pictogram must have a **key** to show what each picture represents. |

Example   ⚲ represents 2 children

⚲ represents 1 child

⚲ ⚲ ⚲ represents 2 + 2 + 1 = 5 children

1   Janet has done a survey to find the favourite summer sport of pupils in 9M. She has drawn a pictogram.

  a How many members of 9M chose tennis?

  b How many chose swimming?

  c Seven members of 9M chose cricket.
  Draw the missing line for Janet's pictogram in your exercise book.

Favourite summer sport for 9M

Swimming   ⚲ ⚲ ⚲ ⚲ ⚲ ⚲

Tennis   ⚲ ⚲ ⚲

Cricket

Athletics   ⚲ ⚲ ⚲ ⚲

Key: ⚲ represents 2 children

2 Sue has done a survey of pupils in 9P.
She asked them their favourite flavour of sweets.
Here are her results:

Flavour	Strawberry	Lime	Lemon	Orange	Blackcurrant
Number of people	8	4	3	10	5

Sue has drawn a pictogram.
a Sue's pictogram is not very good.
Write down two things that are wrong with it.

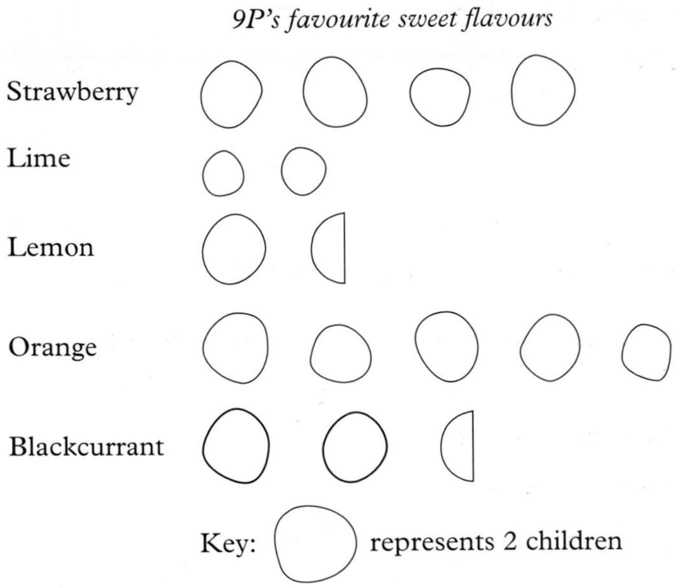

9P's favourite sweet flavours

Strawberry

Lime

Lemon

Orange

Blackcurrant

Key: represents 2 children

b Draw an accurate pictogram for Sue's results.

You may have to *estimate* the number represented by the picture in a pictogram.

Example represents 10 people.

How many people are represented by  ?

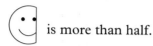 is more than half.

An estimate is 7 people.

3 represents 10 people.

Estimate the number of people represented by:

a   **b**   **c**

4 represents five ice creams.

Estimate the number of ice creams represented by:

a   **b**   **c**

5 represents 10 cats.

a Estimate the number of cats represented by:

(1)   (2)   (3)

b Draw a picture to represent nine cats.
Use the same scale.

6 Some pupils in Year 9 keep pets.
Simon is drawing a pictogram to show them.

Key: 1 picture represents 10 pets

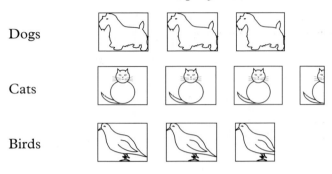

Pets kept by Year 9

Dogs

Cats

Birds

Fish

a How many dogs are kept by Year 9?
b How many cats are kept by Year 9?
c Sally says there are 25 birds.
Explain why Sally is wrong.
d There are 42 fish.
Draw suitable pictures for the fish in your exercise book.

7 Stanthorne High has a drinks machine.
The table shows the different drinks sold during one day.

Flavour of drink	Tea	Coffee	Chocolate	Chicken Soup
Number sold	40	55	23	38

a Draw a pictogram to show the number of drinks sold.

Use ⊔ to represent 10 drinks.

b Give your pictogram the title 'Drinks sold during one day'.
c Give your pictogram a key.

Exercise 4:2

1 Janet has collected data on the number of hours her friends spend on homework in one week.
She has drawn a bar-chart to show her data.

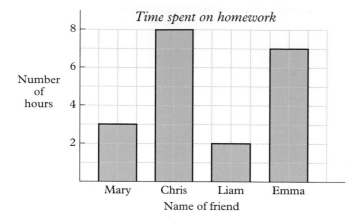

a Mary spent 3 hours on homework.
How many hours did Emma spend?
b Who spent the least time on homework?
c Who spent the most time on homework?

2 This bar-chart shows the time each member of the Williams family spend getting to work or school.

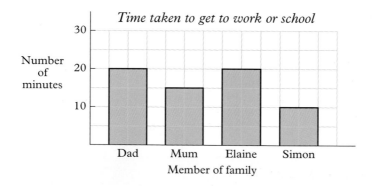

a Who takes the same time to travel as Mr Williams?
b Mr Williams travels 18 miles to work.
Elaine travels just over one mile.
Give a reason why you think they take the same time.
c How long does Simon take to walk to school?
d Elaine and Simon both walk to school.
Does the bar–chart tell you who has to walk the shorter distance?

3 A rental company is finding out how many televisions different families have.
This chart shows the data collected in one estate.

Number of televisions	Number of houses
0	2
1	12
2	8
3	5
4	1

a Yukari has drawn a bar-chart for the data.
Part of Yukari's bar-chart is shown.
Copy the chart.
Use the data to finish it.

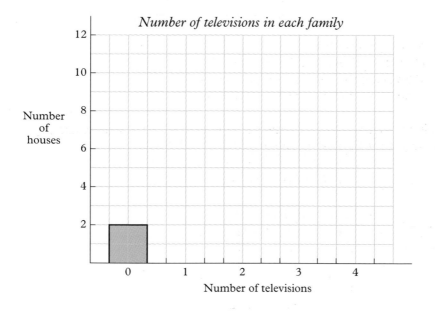

b How many houses had 2 televisions?
c How many houses had more than 2 televisions?
d How many houses were included in the survey?
e How many houses had less than 3 televisions?

4 These bar-charts are for two families.
They show how the families spent money on food one weekend.

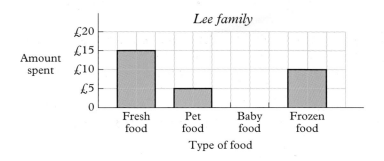

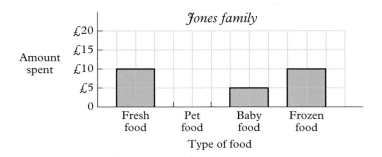

   a Describe one difference between the families.
   b Which family spent the most on fresh food?
   c How much did the Lee family spend on frozen food?
   d How much did the Jones family spend altogether?

5 Paul earns £15 a week delivering papers.
This is how he spent this money last week:

Draw a bar-chart using Paul's data.

2 Tally-tables and groups

Peter and Jane are doing a survey of Year 9.
They are asking pupils to choose a school trip.
They are using a tally-table to help them record the data.

Tally-marks	**Tally-marks** are done in groups of five. The fifth tally-mark goes across the other four: ⦀⦀ This makes them easier to count.

Exercise 4:3

1 Here are the results of Peter and Jane's survey.
They asked pupils to choose a school trip.

Club	Tally	Total
seaside	⦀⦀ ⦀⦀ ⦀⦀	
theatre	⦀⦀ ⦀⦀ |||	
zoo	⦀⦀ ⦀⦀ ⦀⦀ ⦀⦀		
safari park	⦀⦀ ⦀⦀ ⦀⦀ ||||	

a Write down the total for:
   (1) seaside   (2) theatre   (3) zoo   (4) safari park
b How many pupils did Peter and Jane ask?
c Draw a bar-chart to show Peter and Jane's data.

Frequency   **Frequency** means the number of times something happens.

---

2 Ann has done a survey on the number of times people go to the cinema in one month.
She asked 40 people.
This is her data:

2	0	4	5	1	2	1	0	3	3
3	3	2	4	1	0	0	5	4	2
1	4	5	2	4	1	3	0	0	2
4	0	1	1	3	2	2	4	2	2

Copy the tally-table and fill it in.

Number of times	Tally	Frequency
0		
1		
2		
3		
4		
5		

3 Fiona is doing her Food Studies homework.
She has to find out what type of fruit her friends have eaten in the last week.
Here is the data she collected:

apple	pear	banana	apple	pear
grapes	apple	apple	pear	banana
apple	grapes	banana	banana	pear
apple	banana	apple	pear	apple

a Copy this tally-table to record her results.
Fill it in.

Type of fruit	Tally	Frequency
apple		
pear		
banana		
grapes		

b Draw a pictogram to show Fiona's data.
Use one picture to represent two fruits.

Example　Here are the times the pupils in 9R took to get to school this morning. The times are in minutes, correct to the nearest minute.

16	4	7	23	11	9	29	31	26	33
21	33	17	4	19	26	20	11	7	27
5	16	14	5	26	13	19	37	20	6

a Make a tally-table of the data.
b Draw a bar-chart.

a A tally-table showing every minute separately would be too long. This sort of data needs to be in groups.

Time (min)	Tally	Frequency
1–10	ＮＩ ＩＩＩ	8
11–20	ＮＩ ＮＩ Ｉ	11
21–30	ＮＩ ＩＩ	7
31–40	ＩＩＩＩ	4

b When you use grouped data the bars must touch.

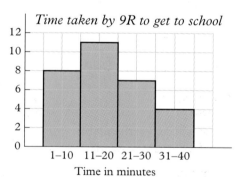

Time taken by 9R to get to school

Frequency

Time in minutes

4 Mrs Grant asked her class how many minutes they had spent on their English homework.
Here is the data she collected:

15	38	53	44	36	73	40	14	25	28
67	51	30	43	28	19	28	36	44	64
64	52	57	38	42	51	17	23	25	77

Copy the tally-table and fill it in.

Time (min)	Tally	Frequency
0–19		
20–39		
40–59		
60–79		

16 17 18 19 20 21

5 Alex organised a quiz for her youth club.
The scores were out of 30.
Here are the scores of 32 people who took part:

16	23	28	3	18	21	26	22
29	27	14	17	10	26	27	20
21	25	16	8	29	23	17	11
28	29	18	27	4	8	13	23

a Copy the tally-table.
Fill it in.

Number of marks	Tally	Total
1–5		
6–10		
11–15		
16–20		
21–25		
26–30		

b How many people scored marks in the range 6–10?
c Which range of marks did most people score?
d Was Alex's quiz an easy quiz or a hard one?
e Give a reason for your answer to part **d**.
f Copy these axes.
Draw a bar-chart.

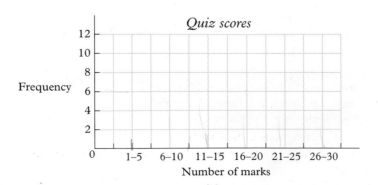

6 A market gardener decided to test two types of tomato plant.
He kept a record of the weights of all the tomatoes from each plant.
He used his data to draw these two bar-charts:

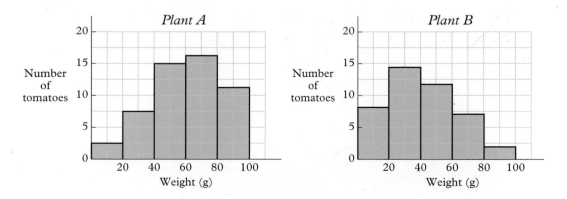

a Which plant gave more heavy tomatoes?
b A restaurant wants to buy small tomatoes.
Which plant should the market gardener use to produce small tomatoes ?

7 Terry's bar-chart shows the height of some seedlings.
Terry measured the plants two months after the seeds were sown.

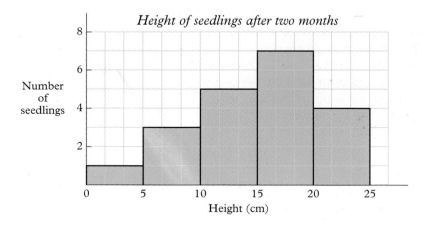

a How many plants are more than 20 cm tall?
b How many plants are less than 10 cm tall?
c How many plants were measured?
d The largest group has seven plants.
Jane said that these plants must be the tallest as they have the highest bar.
Is Jane right? Explain your answer.

3 Pie-charts

Below sea level

At sea level

Above sea level

The land in Holland is mostly flat.
The small number of hills are not very high.
The pie-chart shows how the land is divided between below sea level, at sea level, and above sea level.

Exercise 4:4

1 **a** Use compasses to draw a
circle with radius 4 cm.
Cut the circle out.
b Fold the circle in half.
Fold the circle again into quarters.
Then unfold your circle.
c Colour half the circle red.
Colour a quarter of the circle blue.
Colour the other quarter yellow.

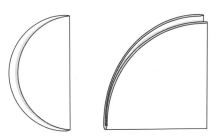

2 The whole circle is 100%.
a What is half of 100%?
b What percentage of the circle is red?
c What percentage is green?
d What percentage is yellow?

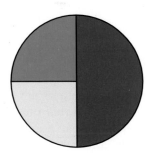

You may have to *estimate* the percentage shown by 'slices' of pie-charts.

Example

This pie-chart shows what pupils in a school do for lunch.

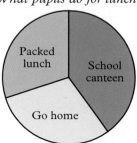

What pupils do for lunch

a Estimate the percentage that buy lunch in the school canteen.

b Estimate the percentage that bring a packed lunch.

a The 'slice' for school canteen is a bit less than half.
It is a bit less than 50%.
An estimate is 40%.

b The 'slice' for packed lunch is a bit more than a quarter.
It is a bit more than 25%.
An estimate is 30%.

3 Estimate the percentage that is coloured red in each of these pie-charts.

a

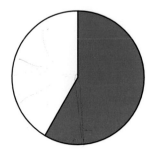

c

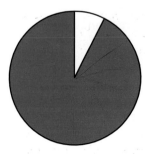

b

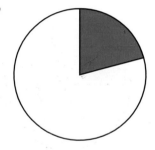

d

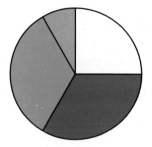

4 These pie-charts show how pupils in different years feel about school uniform.

Year 7   *Year 9*   *Year 11*

   a Which year has about 20% who like uniform?
   b Which year has about 60% who like uniform?
   c Which year has the largest percentage of pupils who like wearing uniform?
      Estimate the percentage for this year.

Exercise 4:5

1 **a** What percentage is yellow?
   b What fraction is yellow?
   c What percentage is red?
   d What fraction is red?
   e The pie-chart represents 12 coloured counters.
      How many of the counters are yellow?
   f How many of the counters are red?

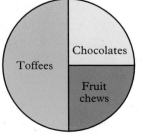

2 The contents of a box of mixed sweets is shown in this pie-chart.
   a What percentage are fruit chews?
   b What fraction are toffees?
   c The box contains 60 sweets.
      How many toffees are there?
   d How many chocolates are there?

Contents of box of mixed sweets

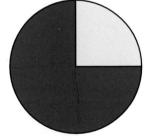

3 This pie-chart shows what cornflakes are made of.

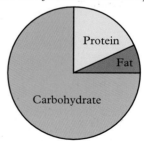
What cornflakes are made up of

   a What fraction is carbohydrate?
   b About what percentage is protein?
● **c** David eats 200 g of cornflakes for breakfast.
     What weight of carbohydrate is this?

4 This pie-chart shows the contents of a type of cheese.

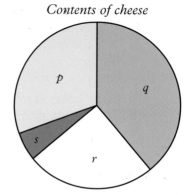
Contents of cheese

Contents	Percentage
fat	31%
water	38%
protein	25%
carbohydrate	

   a Work out the percentage that is carbohydrate.
   b Why must part *s* represent carbohydrate?
   c Write down what part *r* in the pie-chart represents.
   d Write down what part *p* represents.

5 This pie-chart shows how Michael spent 24 hours last weekend.
   a What fraction of his time did he spend eating and helping at home?
   b How many hours did he spend eating and helping at home?
   c He spent about the same length of time on another activity. What was it?
   d Estimate the percentage of his time that he spent sleeping.

How Michael spent 24 hours

1   The pictogram shows fruit bought in the school canteen in one week.
Each picture in the pictogram represents 10 fruits.

Fruit sold in one week

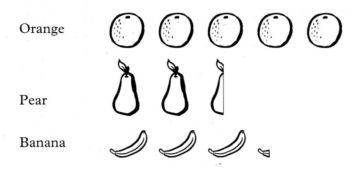

Orange

Pear

Banana

Apple

a   How many oranges were sold?
b   How many pears were sold?
c   John said 35 bananas were sold.
Explain why John is wrong.
d   48 apples were sold.
Draw the pictures for the apples in your exercise book.

2   Hitesh is doing a survey on the use of bicycles.
He has asked each member of the class how many bicycles their family owns.
Here are his results:

3	2	1	3	0	1	3	2	3	2
2	3	2	2	3	2	5	3	0	3
1	0	3	3	2	3	1	2	1	3

a   Copy this tally-table.
Tally the results.

Number of bicycles	Tally	Total
0		
1		
2		
3		
4		
5		

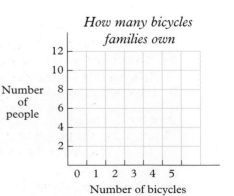

How many bicycles families own

b   Copy the axes.
Draw a bar-chart of the survey.

3 The bar-chart shows
the number of people
who use a snack bar
before lunch one
Monday.

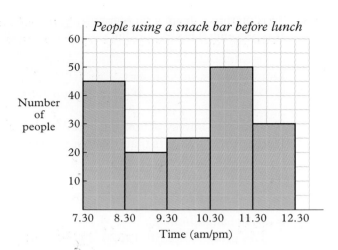

People using a snack bar before lunch

Number
of
people

Time (am/pm)

a What time does the snack bar open?
b How many people use the snack bar during the first hour?
c Why do you think that the people use the snack bar at this time?
d The snack bar has another busy hour.
When is it busy again?
e Why is the snack bar busy for a second time?

4 These pie-charts show the sales of drinks in a cafe.
One is for a day in January and one is for a day in July.

January

Cold
drinks

Coffee

Tea

July

Coffee

Tea

Cold
drinks

For the day in January:
a Estimate the percentage of sales that are coffee.
b What fraction of sales are cold drinks?
c The takings for this day were £184.
How much money was spent on cold drinks?

For the day in July:
d Estimate the percentage of cold drink sales.
Give a reason for the increase in sales of cold drinks between
January and July.
e The takings for this day were £210.
About how much money was spent on coffee and tea together?

1 This picture is from a pictogram.
It represents 8 hours sunshine.

Draw pictures to represent:
a 4 hours sunshine **b** 2 hours sunshine **c** 7 hours sunshine

2 Naveed did a survey of how much 9M spent on snacks at morning break.
Here are the amounts that 9M spent in pence.

0	20	10	40	32	25	35	55	37	10
29	33	30	25	15	21	27	8	25	48
45	35	15	35	20	50	26	31	40	20

a Make a tally-table.
Use groups 0–9, 10–19, 20–29, 30–39, 40–49, 50–59.

b Draw a bar-chart to
show the data.
Make the money axis
look like this:

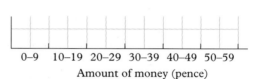

0–9   10–19 20–29 30–39 40–49 50–59
Amount of money (pence)

3 Jean did a traffic survey outside her school.
She recorded how many people were in each car.
These pie-charts show her data for two times of the day.
Each pie-chart represents 60 cars.

Between 8.15 am and 9.00 am   *Between 9.00 am and 9.45 am*

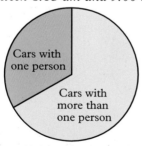

Cars with one person / Cars with more than one person

a Estimate the percentage of cars with just one person between
8.15 am and 9.00 am.
b Estimate the percentage of cars with just one person between
9.00 am and 9.45 am.
c How did the percentage of cars with one person alter between the
two times?
Can you suggest a reason for this?
d Sketch a new pie-chart showing the percentage of cars with one
person for the total time from 8.15 am to 9.45 am.

You may have to *estimate* the number represented by a picture in a **pictogram**.

Example   represents 10 people.

How many people are represented by ?

 is more than half.

An estimate is 7 people.

Example Here are the times the pupils in 9R took to get to school this morning.
The times are in minutes, correct to the nearest minute.
A **tally-table** showing every minute would be too long.
This sort of data needs to be tallied in groups.

Time (min)	Tally	Frequency
1–10	﹀﹀ |||	8
11–20	﹀﹀ ﹀﹀		11
21–30	﹀﹀ ||	7
31–40	||||	4

When you use grouped data the bars must touch.

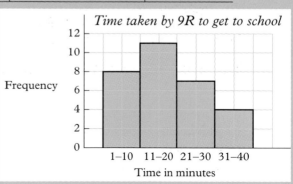

Time taken by 9R to get to school

Example You may have to *estimate* the percentage shown by 'slices' of pie charts.
The 'slice' for school canteen in this **pie-chart** is a bit less than half.
It is a bit less than 50%.
An estimate is 40%.
The 'slice' for packed lunch is a bit more than a quarter.
It is a bit more than 25%.
An estimate is 30%.

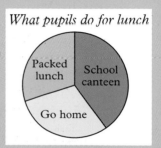

What pupils do for lunch

1   This picture represents 10 houses

Estimate the number of houses represented by:

a   **b**   **c**

2   Some babies were born in a hospital during one weekend.
Here is a bar-chart of their weights:

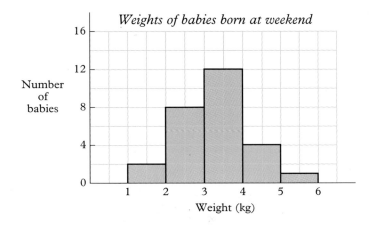

a   How many babies weighed between 3 and 4 kg ?
b   How many babies weighed more than 4 kg ?
c   How many babies weighed more than 3 kg ?
d   How many babies were born at the hospital during the weekend?

3   Penny asked her class where they would like to go on holiday.
She drew this pie-chart to show her results.

Where Penny's class want to go on holiday

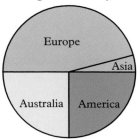

a   What fraction of her class chose Australia?
b   Estimate the percentage of Penny's class that chose Europe.
c   There are 28 pupils in Penny's class. How many chose America?

5 Estimation and measurement

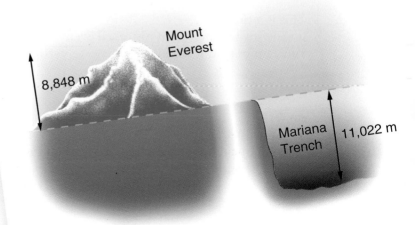

Mount Everest

8,848 m

Mariana Trench

11,022 m

Imagine you could drop Mount Everest into the Mariana Trench in the Pacific Ocean.

How many kilometres of water would there be above the top of the mountain?

1 Estimation

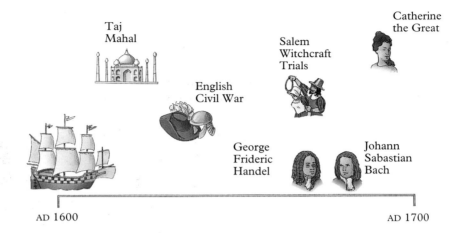

Taj Mahal

English Civil War

Salem Witchcraft Trials

Catherine the Great

George Frideric Handel

Johann Sabastian Bach

AD 1600   AD 1700

This is a time line.
It shows when important historical events happened.
You cannot read it exactly. You have to **estimate** the dates.

Exercise 5:1

1 Estimate the dates of the events marked on these time lines.

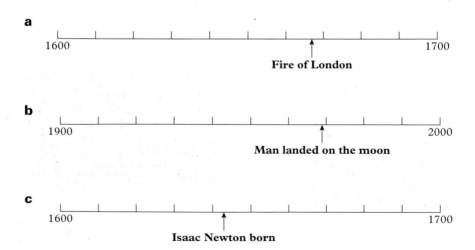

a

1600   1700

Fire of London

b

1900   2000

Man landed on the moon

c

1600   1700

Isaac Newton born

Estimating lengths

You can estimate lengths using kilometres, metres, centimetres or millimetres.
Remember that these are written as **km**, **m**, **cm** or **mm**.

Exercise 5:2

Write down estimates for each of these lengths.
Do not forget to choose **km**, **m**, **cm** or **mm** for each one.

1   **a**

     b

2   **a**

     b

3   **a**

     b

Exercise 5:3

In each of these questions there are two objects.
Write **less than half** if the shorter one is less than half the length of the longer one.
Write **more than half** if it is more than half the length of the longer one.

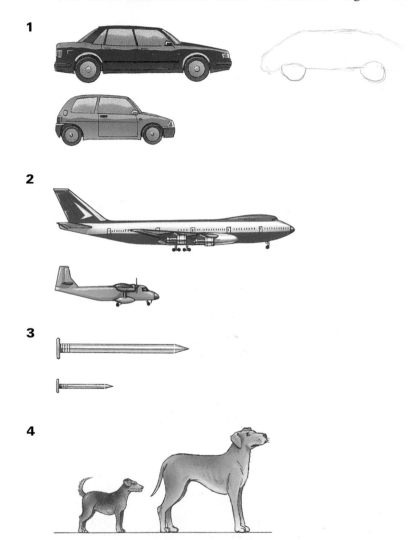

1

2

3

4

5 These are the lengths of the longer object in each picture.
Write down estimates for the length of the shorter object.

    a Car is 6 m   **c** Nail is 5 cm

    b Jumbo Jet is 70 m   **d** Dog is 90 cm

Sometimes you can compare one object with another.

Exercise 5:4

For each question, estimate the lengths marked.
One length is marked in each picture.

1

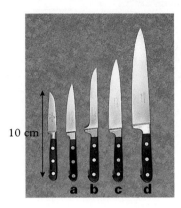

2

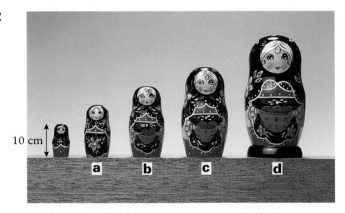

3

Exercise 5:5

There are two objects in each picture.
One is much taller than the other.
Estimate the missing length in each picture.

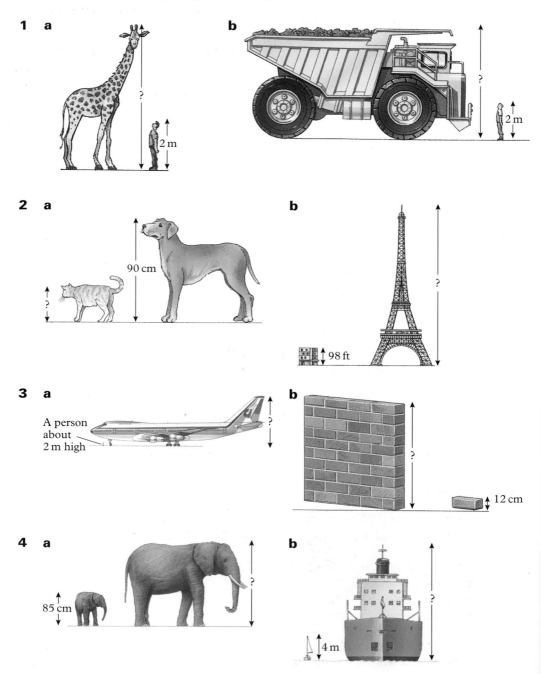

1 a ? 2 m

b ? 2 m

2 a ? 90 cm

b ? 98 ft

3 a A person about 2 m high ?

b ? 12 cm

4 a 85 cm ?

b ? 4 m

2   New for old

The car and the bikes together have a height of 1.95 m.

Only vehicles less than six feet three inches (6' 3") can get into the car park.

Can this car get into the car park?

There are two systems of measurement.

Metric system	Most countries use the **metric system**. It uses metres to measure length. It uses grams to measure weight. 1 metre = 100 cm 1 kilogram = 1000 g
Imperial system	The **imperial system** is older and is used mainly in Britain and the USA. It uses yards, feet and inches for length. It uses stones, pounds and ounces for weight.
Length	1 inch is about 2.5 cm 1 foot (12 inches) is about 30 cm 1 yard (3 feet) is about 90 cm

---

Exercise 5:6

1   Copy these and fill in the gaps.
   a   2 inches is about … cm   **b**   4 inches is about … cm

2   Copy these and fill in the gaps.
   a   1 foot is about … cm
   b   2 feet is about … cm
   c   5 feet is about … cm

3   Andy is 6 feet tall.
   How tall is Andy in centimetres?

4 This diagram shows a full size hockey pitch.
The measurements are in yards.
   a Sketch the pitch in your book.
   b Convert the distances into centimetres.
• **c** Convert your answers to **b** into metres.

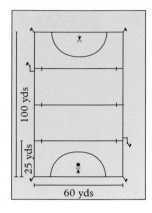

Mile	One **mile** is about 1600 m.

Kilometre	One mile is about 1.6 **kilometres**.

To change miles into kilometres you multiply by 1.6
To change kilometres into miles you divide by 1.6

Example   Convert 20 miles to kilometres.
$20 \times 1.6 = 32$ km

---

5 Convert these distances into kilometres.
   a The distance from Leeds to
       Sheffield is 36 miles.
   b The distance from Manchester to
       London is 200 miles.
   c The distance from Edinburgh to
       Aberdeen is 125 miles.
   d The distance from Lands End to
       John o'Groats is 850 miles.

6 Convert these distances into miles.
   a The distance from Leeds to Nottingham is 120 km.
   b The distance from Exeter to Bristol is 135 km.

Exercise 5:7

Ben Down the PE teacher is clearing out some old cupboards.
He finds some old school records from 1961.

They show all the school athletics records.
They are all in imperial units.

He decides to convert them into metric units.

Remember: 1 yard is about 0.9 m
               1 foot is about 30 cm
               100 cm is the same as 1 m

1   Copy out the records. Convert the distances into metric units.

Event	Boys Record	Girls Record
Under 12 Shot Putt	6 yards	$4\frac{1}{2}$ yards
Under 14 Shot Putt	$7\frac{1}{2}$ yards	6 yards
Under 16 Shot Putt	9 yards	8 yards
Under 18 Shot Putt	12 yards	$10\frac{1}{2}$ yards
Under 12 Javelin	18 yards	19 yards
Under 14 Javelin	25 yards	23 yards
Under 16 Javelin	32 yards	27 yards
Under 18 Javelin	37 yards	30 yards
Under 14 High Jump	42 inches	46 inches
Under 16 High Jump	64 inches	58 inches
Under 14 Long Jump	8 feet	$7\frac{1}{2}$ feet
Under 16 Long Jump	$11\frac{1}{2}$ feet	10 feet

Weight

The metric unit for measuring weight is the gram.

Remember:
   1000 g = 1 kg
   1000 kg = 1 tonne

The imperial units of weight are ounces, pounds and stones.
   16 ounces = 1 pound
   14 pounds = 1 stone

To convert from one to the other:
   1 kg is about 2.2 pounds
   1 ounce is about 30 g
   1 pound is about 450 g

Example

A large box of washing powder weighs 3 kg.
What does it weigh in pounds?

1 kg is about 2.2 pounds
3 kg is about 3 × 2.2 = 6.6 pounds

Exercise 5:8

1   Jo buys a 4 kg bag of peat for her garden.
   What is 4 kg in pounds?

2   Jane buys $\frac{1}{2}$ kg of salt.
   What is $\frac{1}{2}$ kg in pounds?

3   Pardeep buys 2 pounds of minced beef.
   What is 2 pounds in grams?

4   Alan weighs 7 stone.
   a   How many pounds does he weigh? (1 stone = 14 pounds)
   b   Work out Alan's weight in grams. (1 pound is about 450 g)
   c   Work out Alan's weight in kilograms. (1 kg = 1000 g)

5   Work out your own weight in kilograms.

Example

A small cat weighs 5 pounds.
What is 5 pounds in kilograms?

Each 2.2 pounds makes 1 kilogram.
You need to know how many lots of 2.2 there are in 5.
You divide 5 by 2.2:

$$5 \div 2.2 = 2.3 \text{ kg}$$

The cat weighs 2.3 kg.

6 Steven likes ten pin bowling.
The weights of the bowling balls are in pounds.

a Copy this table:

Ball	Weight in pounds	Weight in kg
yellow	8	
red	10	
orange	12	
blue	14	
black	16	

b Work out the weight of each bowling ball in kilograms.

7 An average baby elephant weighs 242 pounds.
Work out the weight of a baby elephant in kilograms.

8 An average human baby when born
weighs about 7 pounds.
Work out the weight of a baby in kilograms.

9 A fully grown elephant weighs around
13 200 pounds.
Work out this weight in kilograms.

● **10** A fully grown woman weighs around
10 stones.
Work out this weight in kilograms.

3 Perimeter

These athletes are running the 1500 m race.

The distance around the inside lane of this track is 400 m.

How many laps of the track will they need to do?

Perimeter

The total distance around the outside edges of a shape is called its **perimeter**.

Example

Find the perimeter of this shape:

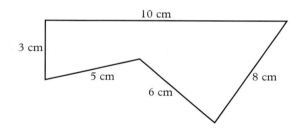

Add up the lengths:
8 cm + 6 cm + 5 cm + 3 cm + 10 cm = 32 cm
The perimeter is 32 cm.

Exercise 5:9

Find the perimeter of each of these shapes:

1

5 cm

3 cm   3 cm

5 cm

2

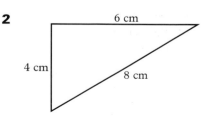

6 cm

4 cm

8 cm

3

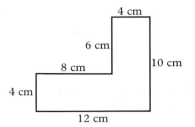

6

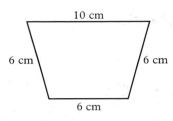

4

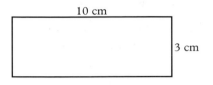

● **7**

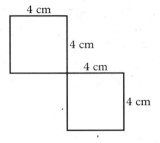

5

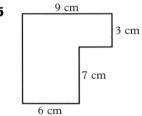

● **8**

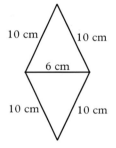

9 A rectangular field measures 60 m by 40 m.
A farmer wants to put a fence around its perimeter.
Find the length of the fence.

● **10** This parcel measures
30 cm by 70 cm by 15 cm.
Val wants to tie it up with
string in both directions.

How much string
will Val need?

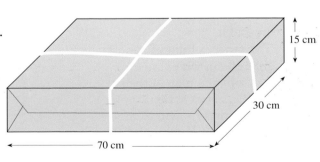

5

| **Convert units** | We often need to **convert** from one metric unit to another. |

10 mm = 1 cm
To change from mm to cm   ÷ 10
To change from cm to mm   × 10

100 cm = 1 m
To change from cm to m   ÷ 100
To change from m to cm   × 100

Examples

1   Convert 630 cm to m.

630 ÷ 100 = 6.3
630 cm = 6.3 m

2   Convert 60 cm to mm.

60 × 10 = 600
60 cm = 600 mm

Exercise 5:10

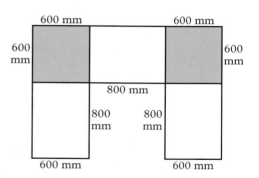

1   Kitchen units are 600 mm deep.
Convert 600 mm into cm.

2   Each of these cupboards is 800 mm wide.
  a   Find the perimeter of the cupboards in mm.
  b   Change your answer to **a** into cm.
  c   Change your answer to **b** into m.

3   A building site needs a fence
around it.
The fence is built of panels which
are 150 cm wide.
  a   Copy the plan of the site.
     It is not to scale.
  b   Change 150 cm to m.
  c   Work out the number of
     panels needed for each length.
  d   How many panels will be
     needed altogether?

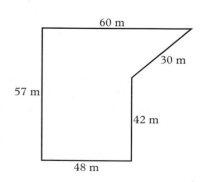

Exercise 5:11

Perimeters and tiles

You will need square tiles for this exercise.

This tile has a perimeter of 4.

You can join two tiles edge to edge like this:
This shape has a perimeter of 6.

You can join them corner to corner like this:
This shape has a perimeter of 8.

You are not allowed to join the tiles in any other way.

With two tiles:
6 is the smallest perimeter you can make,
8 is the biggest perimeter you can make.

1 Use three tiles.
  a Draw the arrangement of three tiles that gives the biggest perimeter.
  b What is the biggest possible perimeter?
  c Draw the arrangement of three tiles that gives the smallest perimeter.
  d What is the smallest possible perimeter?

2 Use four tiles.
  a Draw the arrangement of four tiles that gives the biggest perimeter.
  b What is the biggest possible perimeter?
  c Draw the arrangement of four tiles that gives the smallest perimeter.
  d What is the smallest possible perimeter?

3 Try more tiles. Record your results.
  Look for patterns.
  Write a report on what you find out.

4 You could try other types of tiles.

1 The units on a jar of marmalade are missing.
Copy and fill in:
   a The jar contains 1 ....
   b The jar contains 454 ....
Choose from **kilograms, pounds, ounces, grams**.

2 This is a plan of a bedroom.
   a Estimate the width of the bed.
   b Estimate the length of the bed.
   c Estimate the width of the room.
   d Estimate the length of the room.

3 Look at the girl in this picture.
Copy and fill in.
   a The girl is about 1.62 .... tall.
   b The girl is just over 5 .... tall.
   c The girl weighs about 7............ .
   d The girl weighs about 45 ....
Choose from **stones, pounds,
kilograms, inches, metres, feet**.

4 These shapes are drawn on triangular dotty paper.
The dots are all 1 cm apart.

Find the perimeter of each shape.

a 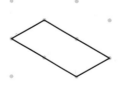   **b**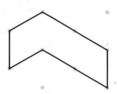

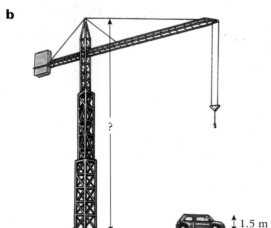

5   There are two objects in each picture.
One is much taller than the other.
Estimate the missing length in each picture.

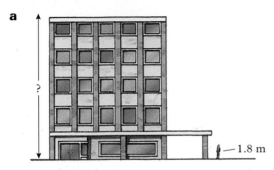

a

— 1.8 m

b

?

1.5 m

6   Copy these and fill in the gaps.
a   3 feet is about .... cm.
b   6 feet is about .... cm.
c   20 feet is about .... cm.
d   Convert your answer to **c** into metres.

1   This is the plan of a basketball court.
All the measurements are in metres.
Make a copy of the plan.
Mark on the distances in feet.

Remember: 1 foot is about 30 cm or 0.3 m
1 metre is about 3.25 feet

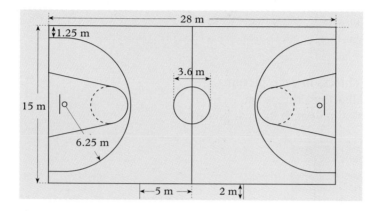

2   This shape has a small area but a big perimeter.
Draw a shape with a bigger area but a smaller perimeter.

3   Find the perimeter of this shape.
All the distances are in centimetres.

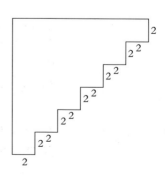

- **Metric system** — The **metric system** uses centimetres and metres to measure length. It uses grams and kilograms to measure weight.

  Imperial system — The **imperial system** is older.
  It uses yards, feet and inches for length.
  It uses stones, pounds and ounces for weight.

  1 inch is about 2.5 cm
  1 foot (12 inches) is about 30 cm
  1 yard (3 feet) is about 90 cm

- **Mile** — One **mile** is about 1600 m.

  Kilometre — This is the same as 1.6 kilometres.
  8 km is about 5 miles

  Example — Convert 20 miles to kilometres.
  $20 \times 1.6 = 32$ km

- **Weight** — The metric unit for measuring weight is the gram.

  Remember:
  1000 g = 1 kg
  1000 kg = 1 tonne

  The imperial units of weight are ounces, pounds and stones.
  16 ounces = 1 pound
  14 pounds = 1 stone

  To convert from one to the other:
  1 kg is about 2.2 pounds
  1 ounce is about 30 g
  1 pound is about 450 g

  Example — A large box of washing powder weighs 3 kg.
  What does it weigh in pounds?

  1 kg is about 2.2 pounds
  3 kg is about 6.6 pounds

- **Perimeter** — The total distance around the outside edges of a shape is called its **perimeter**.

1 Write down estimates for each of these lengths.
Choose **km**, **m**, **cm** or **mm** for each one.

a

b

2 Look at this picture. Estimate the lengths marked **a** to **d**.
One length is marked.

3 Copy these sentences.
Fill in the gaps.
  a Two pounds is about .... grams.
  b 1 metre is just over 3 ....
  c 6 inches is about .... cm.
  d 4 kg is about 9 ....

4 Find the perimeter of these shapes.

a

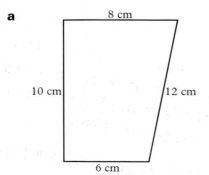

b

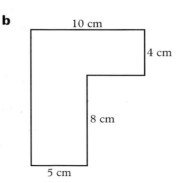

6 Volume

QUESTIONS

EXTENSION

SUMMARY

TEST YOURSELF

Many scientists believe that global warming is melting the polar ice caps.

A study by the Inter-governmental Panel for Climate Change estimates that by the year 2070 global sea level could be as much as 71 cm higher than it is now.

Find out which parts of the UK would be underwater.

1 Units of capacity

Concrete is used to make the foundations of a house.

The builder needs to know how much concrete to mix.

He works out the volume of the hole that he needs to fill.

There is a scale on the side of this jug.
It tells you how many cups of coffee the jug holds.

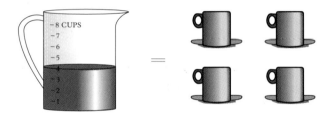

The jug will fill four cups with coffee.

Exercise 6:1

1 Chizuru has made coffee for her friends.
   a How many cups of coffee has she made?

   b She fills some cups with coffee.
   The jug now looks like this:
   How many cups has she filled?

2 This jug of coffee will
fill three cups.

How many cups will these jugs fill?

a  **b**  **c**

3 Peter has put five cups of water into this jug.
How many cups should be written on the
line marked by the arrow?

4 John has put five cups of water into this jug:
How many cups should be written on the
line marked with the arrow?

The doctor is giving medicine to some children.
She gives one 5 ml spoonful to each child.

This container holds enough medicine for three children.

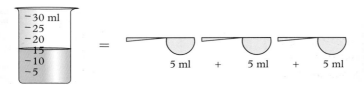

Exercise 6:2

A doctor is giving medicine to some children. She gives one 5 ml spoonful to each child.

1 How many children can the doctor treat with these?

a

```
─ 30 ml
─ 25
─ 20
─ 15
─ 10
─ 5
```

b

```
─ 30 ml
─ 25
─ 20
─ 15
─ 10
─ 5
```

c

```
─ 30 ml
─ 25
─ 20
─ 15
─ 10
─ 5
```

2 This bottle of cough mixture contains 40 ml.
   a How many 5 ml spoons will it fill?
   b Gemma has to take one 5 ml spoonful of cough mixture twice a day. How many days will the bottle last?

Cough
Mixture
40 ml

3 A factory makes solid silver medals.
Each medal needs 10 ml of molten silver.

How many medals can you make from:

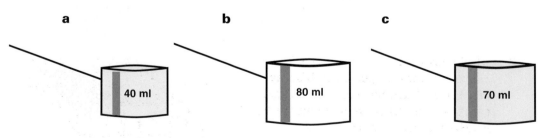

a

40 ml

b

80 ml

c

70 ml

d 20 ml of molten silver   **e** 100 ml of molten silver?

2   How many?

Amy and Rhian are organising a stall for the school autumn fair.

Exercise 6:3

1   **a**   Amy decides to stack the cans like this:
How many cans does she need?

b   Rhian wants to stack the cans this way:
Does she use the same number of cans?

c   Which two arrangements use the same number of cans?

(1)    (2)    (3)

2   Afzal works each Saturday on a farm.
He has to put the eggs from this tray
into boxes. Each box contains 6 eggs.
a   How many eggs are there on the tray?
b   How many boxes can he fill?

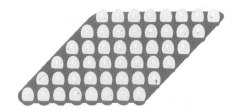

3 a How many apples are there on this tray?

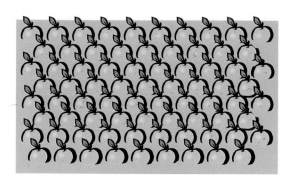

b How many packets of 4 apples like this can you make?

4 How many Toblerones are there?

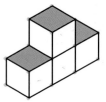

Exercise 6:4

You will need some cubes for this exercise.
Use the cubes to make the shape in each picture.
Write down the number of cubes that you use for each shape.

1

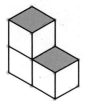

2

3

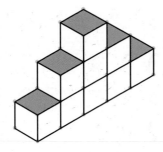

6

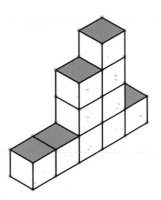

4

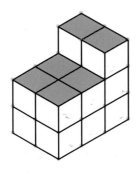

7

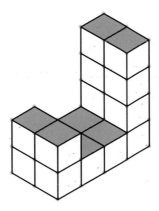

5

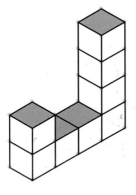

8

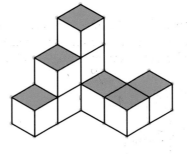

3   Volume of a cuboid

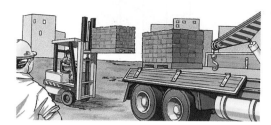

The builder needs to know how
many bricks there are in each
pack.

Exercise 6:5

Work out how many bricks there are in each of these packs.

1   **2**   **3**

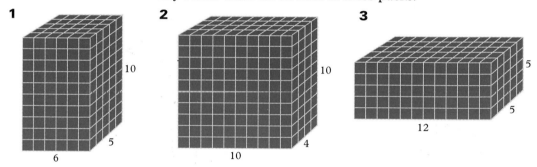

When you need to know a volume you can use cm³.

A cube that has sides of 1 cm is
called a 1 cm cube.
You say that it has a volume of 1 cm cubed.
You write this as 1 cm³.

This shape is made from 2 cubes.
Its volume is 2 cm³.

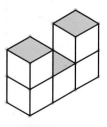

This shape is made from 5 cubes.
Its volume is 5 cm³.

Exercise 6:6

Jenny made these shapes with 1 cm cubes.
What is the volume of each shape?

1

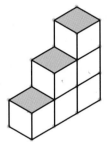

4

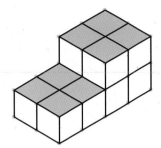

2

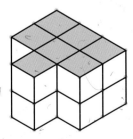

5

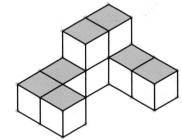

3

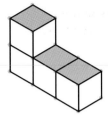

6

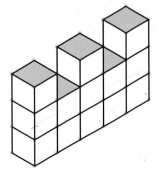

Look at this block of 1 cm cubes.
There is a fast way to find the volume.

Count the number of cubes along the
length, width and height.

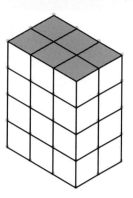

Volume = length × width × height
     = 3 × 2 × 4
     = 24 cm³

Exercise 6:7

Find the volume of these blocks of 1 cm cubes.
Write your answers in cm³.

1

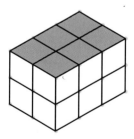

3

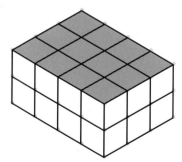

2

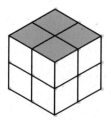

4

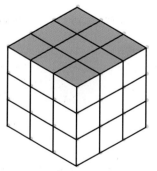

5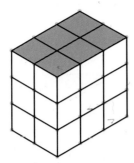

6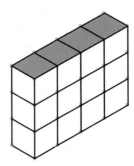

The blocks of cubes in the last exercise are all in the shape of cuboids.
You can measure the length of each side of the cuboid instead of
counting cubes.
These lengths are used to calculate the volume.

**Volume of
a cuboid**

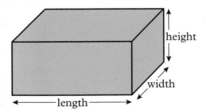

Volume of a cuboid = length × width × height

Example   Find the volume of this cuboid.

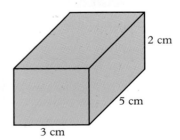

Volume = length × width × height
$$= 3 \times 5 \times 2$$
$$= 30 \text{ cm}^3$$

Exercise 6:8

Work out the volume of these cuboids.

1

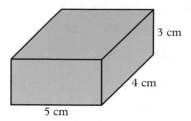

3 cm
4 cm
5 cm

4

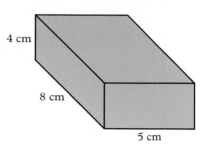

4 cm
8 cm
5 cm

2

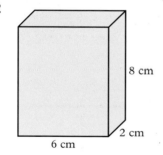

8 cm
2 cm
6 cm

5

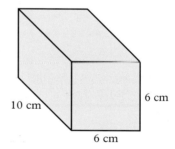

6 cm
10 cm
6 cm

3

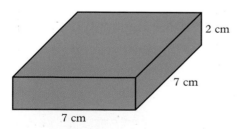

2 cm
7 cm
7 cm

6

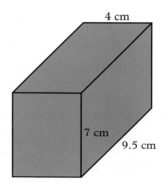

4 cm
7 cm
9.5 cm

Exercise 6:9

Work out the volume of these containers.

1

15 cm
10 cm
30 cm

4

40 cm
21 cm
7 cm

2

10 cm
5 cm
10 cm

5

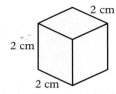

2 cm
8 cm
18 cm

3

15 cm
9 cm
27 cm

6

10 cm
8 cm
2 cm

● **7** **a** Work out the volume of this cube.
     b What is the volume of four of these cubes?

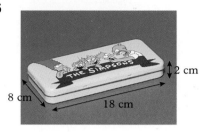

2 cm
2 cm
2 cm

  c What is the volume of this box?
  d Which has the greater volume,
        four of the cubes or the box?
  e Will the cubes fit into the box?
        Explain your answer.

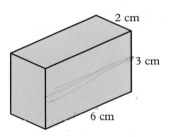

2 cm
3 cm
6 cm

Design a box

The Happy Cow Cheese Company wants a box for 12 Happy Cheeses.

Each of the cheeses is a cube of side 2 cm.

You will need some cubes for this investigation.

1   Arrange the 12 cheeses to make a cuboid.

2   Find the volume of the cuboid.

3   Make a different cuboid with the 12 cheeses.

4   Does the new cuboid have the same volume as the cuboid you made first?

5   How many different cuboids can you make using the 12 cheeses?
Do they all have the same volume?

6   Which arrangement of the 12 cheeses do you think the Happy Cow Cheese
Company will choose?
Explain your reasons.

7   The pack of 12 cubes is to be a standard size.
The company wants to sell different sizes of packs of Happy Cheese cubes.
They want different numbers of cubes in the different sizes.
Design some packs containing different numbers of cubes.

1 How many cups of coffee does each jug contain?

a  **b**   **c**

2 A bottle of cough mixture contains 100 ml.
  a How many 5 ml spoons will it fill?
  b Toby has to take four 5 ml spoonfuls of cough mixture each day.
     How many days will the bottle last?

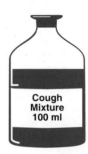

3 This 1 litre bottle fills 6 glasses.
  a How many glasses will 3 bottles fill?
  b How many bottles are needed to fill 12 glasses?

4 This is a can of condensed soup.
   You have to add 2 cans of water to make the soup.
   How many millilitres of soup will you have?

5 This water butt contains 250 litres of water.
   A watering can holds 5 litres of water.
   How many times can you fill the can from the water butt?

6 This carton contains 500 ml of orange juice.
   It is emptied into 5 glasses.
   How much orange juice is in each glass?

7 Ben is washing cars to raise money for charity.
This is the car shampoo that he uses.
1 capful of shampoo is needed to wash one car.

a How many cars can Ben wash with this
bottle of shampoo?

b This is the bottle after Ben spent
one evening washing cars.
How many cars did he wash?

c Ben washed another seven cars the next evening.
Which of these bottles shows how much shampoo is left?

A   B   C   D

8 Work out the volume of these cuboids.

a

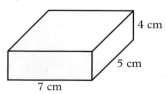

4 cm
5 cm
7 cm

c

10 cm
12 cm
11 cm

b

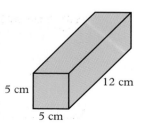

5 cm
12 cm
5 cm

d

12 cm
4 cm
18 cm

1 This beaker is full. It contains 100 ml of water.

Which of these beakers contains 40 ml?

a   **b**   **c**   **d**

2 Jodie can buy oil in these sizes.

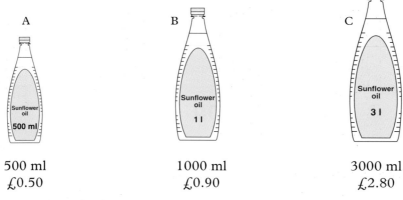

A   B   C

500 ml   1000 ml   3000 ml
£0.50   £0.90   £2.80

  a How much does 100 ml cost for size A?
  b How much does 100 ml cost for size B?
  c How much does 100 ml cost for size C?
  d Which size gives the best value for money?

3 This full-size jar contains 350 ml of honey.

The small portion of honey contains 25 ml.

How many small portions do you need
to get the same volume as the jar of honey?

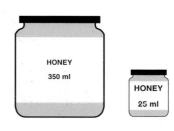

4 **a** Work out the volume of
this packet of butter.

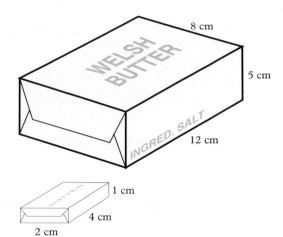

8 cm

5 cm

12 cm

b Work out the volume of
this portion of butter.

1 cm

4 cm

2 cm

c How many small portions do you need to get the same volume as
the packet of butter?

5 Work out the volume of a cuboid with length 11 cm,
width 10 cm and height 5 cm.

6 This prize platform is made
with three cuboids.

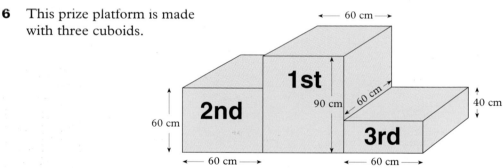

60 cm

1st

2nd

3rd

90 cm

60 cm

60 cm

40 cm

60 cm

60 cm

Work out:
a the volume of each of the three cuboids,
b the total volume of the platform.

7 The volume of this gift box is 240 cm³.
Find the length of the box.

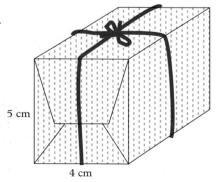

5 cm

4 cm

- This container holds enough medicine for three spoonfuls.

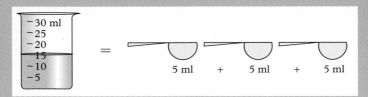

- A cube that has sides of 1 cm is
called a 1 cm cube.
You say that it has a volume of 1 cm cubed.
You write this as 1 cm³.

This shape is made from 2 cubes.
Its volume is 2 cm³.

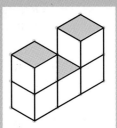

This shape is made from 5 cubes.
Its volume is 5 cm³

- **Volume of a cuboid = length × width × height**

Example   Work out the volume of this
block of 1 cm cubes.

Volume = length × width × height
$$= 3 \times 2 \times 4$$
$$= 24 \text{ cm}^3$$

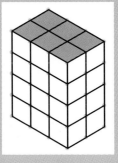

Volume of a cuboid = length × width × height

Example   Find the volume of this cuboid.

Volume = length × width × height
$$= 3 \times 5 \times 2$$
$$= 30 \text{ cm}^3$$

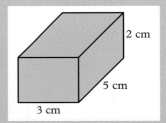

1 This carton contains 800 ml of orange juice.

The glass holds 100 ml.
How many glasses will the carton fill?

100 ml

2 How many cubes are used to make each of these shapes?

a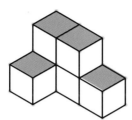

b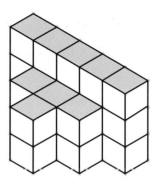

3 These shapes are made with 1 cm cubes.
What is the volume of each shape?

a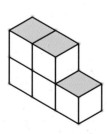

b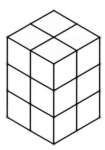

4 Work out the volume of these cuboids.

a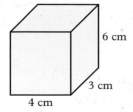

6 cm

3 cm

4 cm

b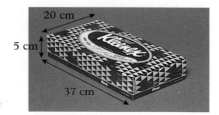

20 cm

5 cm

37 cm

7 Decimals

In printed maths we don't use commas in numbers. Why is this?

1   Getting things in order

This is a picture of the final of the Olympic 100 m race in Atlanta in 1996. The times of the runners are given to the nearest one hundredth of a second. This accuracy is needed to separate the times of the runners.

These are the times in seconds for all of the runners:

9.84   9.89   9.90   9.99   10.00   10.14   10.16

The smallest time is 9.84 s.
Donnovan Bailey won the race because he took the least time.
The next smallest time is 9.89 s. Frank Fredericks came second with this time.
The biggest time is 10.16 s for Michael Green who came last in the race.

Exercise 7:1

1   Copy the times in order.
Start with the smallest.

2   How many seconds are there between first and second?
$9.89 - 9.84 = \ldots$

3   How many seconds are there between second and third?
$\ldots - 9.89 = \ldots$

4   How many seconds are there between first and last?
$\ldots - \ldots = \ldots$

5   **a**   How fast can you run 100 m?
b   How many seconds behind Donnovan Bailey would you be?

Examples   **1**   12.96 is smaller than 13.14 because **12** is smaller than **13**.

   2   14.36 is smaller than 14.61 because **3** is smaller than **6**.

   3   14.26 is smaller than than 14.28 because **6** is smaller than **8**.

Exercise 7.2

1   These are the times in seconds for the men's 200 m final.

A. Boldon 19.80   F. Fredericks 19.68   I. Garcia 20.21
M. Johnson 19.32   M. Marsh 20.48   P. Stevens 20.27
O. Thompson 20.14   J. Williams 20.17

a What is the winning time?
b What is the time for second place?
c List all the times in order.
   Start with the smallest.
d How many seconds are there between first and second?
e How many seconds are there between first and last?

2   These are the times in seconds for the women's 200 m final.

J. Cuthbert 22.60   C. Guidry 22.61   G. Malchugina 22.45
I. Miller 22.41   M. Onyali 22.38   M. Otley 22.24
M-J Perec 22.12   C. Stirrup 22.54

a What is the winning time?
b What is the time for second place?
c List all the times in order.
   Start with the smallest.
d How many seconds are there between first and third?
e How many seconds are there between first and last?

3 These are the distances in metres for the women's long jump final.
The longest jump wins the competition.

C. Ajunawa 7.12   J. Joyner-Kersee 7.00   I. Prandzheva 6.82
N. Boegman 6.73   A. Karczmarek 6.90   T. Vaszi 6.60
C. Brunner 6.49   F. May 7.02   N. Xanthou 6.97
I. Chekhovtsova 6.97   V. Patoulidou 6.37

a What is the winning distance?
b What is the distance for second place?
c List all the distances in order.
     Start with the largest.
d How many metres are there between first and second?
e How many metres are there between first and last?

4 These are the distances in metres for the men's long jump final.

E. Bangue 8.19   G. Huang 7.99   I. Pedroso 7.75
J. Beckford 8.29   A. Ignatov 7.83   M. Powell 8.17
G. Cankar 8.11   C. Lewis 8.50   M. Sunneborn 8.06
A. Glovatskiy 8.07   Y. Naumkin 7.96
J. Greene 8.24   E. Nijs 7.72

a What is the winning distance?
b What is the distance for second place?
c List all the distances in order.
     Start with the largest.
d How many metres are there between first and second?
e How many metres are there between first and third?
f How many metres are there between first and last?

You don't always have 2 dp when you have decimals.
You need to be able to put all decimals in order.

To put decimals in order of size:

1 Look at the number before the decimal point first.

e.g. 3.12 is smaller than 4.34
because 3 is smaller than 4

2 If the numbers before the decimal point are the same
then look at the first numbers after the decimal point.

e.g. 3.463 is smaller than 3.58
because 4 is smaller than 5

3 Sometimes you need to look at the second numbers after the decimal point.

e.g. 3.441 is smaller than 3.45
because the 4 is smaller than the 5

4 Carry on like this. Look at one decimal place at a time.

Exercise 7:3

In questions **1** to **12**, put the numbers in order of size.
Start with the smallest.

1	3.4 5.5	**3**	6.58 4.39	**5**	4.69 4.28	**7**	8.163 8.168
2	4.76 7.46	**4**	3.46 3.58	**6**	7.47 7.49	**8**	6.437 6.432

9 3.46 4.72 4.57   ● **11** 2.465 4.723 4.701

10 8.37 8.45 5.31   ● **12** 3.378 3.356 5.31

● **13** Year 9 have a discus competition. These are the five longest throws.

8.78 m   8.5 m   8.14 m   9.3 m   9.287 m

a Put these distances in order of size. Start with the biggest.
b Which of these distances would be very hard to measure?

2 Working with decimals

The lap times for Formula One races are given to 2 dp. A very small amount of time saved on one lap can make a big difference overall. Drivers look at their lap times to see how they did during the race.

They add their lap times to get their total time for the race. They subtract their lap times to see how much faster they went on their best laps.

Adding decimals without a calculator

You set out additions in columns. This is the same as you do for whole number questions.

Make sure that you put the decimal points underneath each other.

Examples   **1**   Work out 3.35 + 4.3

$$
\begin{array}{r}
3.35 \\
+\ 4.30 \\
\hline
7.65 \\
\end{array}
$$

   ← You can fill the gap with a 0.
Add each column to get the answer.

2   Work out 63.36 + 24.9

$$\begin{array}{r} 63.36 \\ +\ 24.90 \\ \hline .\ 6 \end{array}$$ First add the **6** and the **0** which is there to fill the gap.

$$\begin{array}{r} 63.36 \\ +\ 24.90 \\ \hline .26 \\ \tiny 1 \end{array}$$ Now add the **3** and the **9**. The answer is 12. Put the **2** in the column and carry the **1**.

$$\begin{array}{r} 63.36 \\ +\ 24.90 \\ \hline 8.26 \\ \tiny 1 \end{array}$$ Now add the **3**, the **4** and the **1** that you carried.

$$\begin{array}{r} 63.36 \\ +\ 24.90 \\ \hline 88.26 \\ \tiny 1 \end{array}$$ Finally, add the **6** and the **2**.

Exercise 7:4

Work these out.

1	5.34 + 4.40	**4**	7.46 + 1.82	**7**	72.46 + 19.4
2	2.81 + 5.14	**5**	4.26 + 2.7	**8**	46.5 + 66.37
3	4.26 + 3.12	**6**	6.35 + 3.8	**9**	36.35 + 8.67

10   5.2 + 3.32   **12**   23.2 + 8.3   **14**   3.24 + 23.5

11   45.31 + 22.3   **13**   46.9 + 7.7   **15**   24.2 + 1.9

16   These are 4 lap times in seconds. Find the total time for these laps.

168.34   169.63   167.89   170.12

Subtracting decimals without a calculator

You set out subtractions in columns. This is the same as you do whole number questions.

Make sure that you put the decimal points underneath each other.

Examples

1 Work out 34.73 − 23.31

```
   34.73
−  23.31     Subtract each column to get the answer.
   11.42
```

2 Work out 54.25 − 33.44

```
   54.25     First do 5 − 4
−  33.44
    . 1
```

```
   5⁴.25     You can't take 4 from 2. Borrow one from
−  33.44     the column on the left. The 4 becomes 3
    .81      and the 2 becomes 12. Now do 12 − 4.
```

```
   5⁴.25     Now do 3 − 3 and 5 − 3 to finish the
−  33.44     question.
   20.81
```

Exercise 7:5

Work these out.

1
```
   35.34
−  14.22
```

3
```
   43.74
−  23.43
```

5
```
   53.13
−   1.71
```

2
```
   23.86
−  12.44
```

4
```
   46.38
−   4.77
```

6
```
   26.41
−  24.28
```

In questions **7** to **9** fill in the spaces with a 0.

7
```
   72.4
−  51.25
```

8
```
   34.5
−   6.73
```

9
```
   44.5
−   6.63
```

10 34.35 − 23.24	**12** 56.3 − 23.83	**14** 42.83 − 2.38
11 48.36 − 3.54	**13** 78.2 − 9.39	**15** 82.16 − 23.81

16 These are four lap times in seconds:

168.34   169.63   167.89   170.12

a Which is the fastest time?
b Which is the slowest time?
c How many seconds are there between the fastest and the slowest time?

Multiplying decimals without a calculator

You need to set these out like you do whole number questions. Keep your numbers in columns.

Example   Work out 4.73 × 5

```
  4.73
×    5
     5
     1
```
Start with 5 × 3. The answer is 15.
Put the 5 in the column and carry the 1.

```
  4.73
×    5
   .65
   3  1
```
Now do 5 × 7. The answer is 35.
Add the 1 to get 36.
Put the 6 in the column and carry the 3.
Put the decimal point in.

```
  4.73
×    5
 23.65
   3  1
```
Finally do 5 × 4. The answer is 20.
Add the 3 to get 23.
Put the 3 in the column and put the 2 in the next column.

Exercise 7:6

Work these out.

1   3.31
      $\times$ 3
      ─────

2   5.65
      $\times$ 2
      ─────

3   5.38
      $\times$ 6
      ─────

4 **a** 5.44×3 **d** 4.7×5 **g** 2.3×6
   b 6.46×4 **e** 6.3×6 **h** 3.56×8
   c 8.37×5 **f** 8.3×4 **i** 2.96×7

5 CDs cost £9.85 in a sale.
   How much would you pay for 4 CDs?

6 Alan pays £2.45 each day on the train.
   How much does it cost him for 5 days?

7 Pizzas cost £2.89 each.
   How much do 3 pizzas cost?

8 Gerald makes model planes.
   He sells them for £4.85 each.
   How much does he charge for 5?

You sometimes need to use long multiplication with decimals.
This works like normal long multiplication.
It is important to keep your numbers in columns.

Example   Work out 3.76 × 24

Start with 3.76 × 4

```
     3.76
×       4
   15.04
     3 2
```

Now do 3.76 × 20. Do this by multiplying by 2 first.

```
     3.76
×       2
    7.52
     1 1
```

Then multiply by 10.   7.52 × 10 = 75.2

Now add the two answers together. Fill in missing spaces with 0s.

```
     75.20
×    15.04
    90.24
       1
```

Usually the working looks like this:

```
     3.76
×      24
    75.20
    15.04
    90.24
       1
```

So   3.76 × 24 = 90.24

Exercise 7:7

Work these out.

1	3.75 × 15	**4**	4.75 × 23	**7**	2.93 × 38
2	2.95 × 35	**5**	6.45 × 26	**8**	3.16 × 42
3	6.95 × 25	**6**	8.35 × 36	**9**	8.91 × 46

10 Anne sells clothes on a market stall.
She sells jumpers for £6.45
One day she sells 25 jumpers.
How much money does she get?

11 Bernie sells tapes on a market stall.
A tape costs £4.95
How much money does he get for 28 tapes?

12 Pat makes model toys.
She sells them for £4.75
How much does she get for 34 toys?

13 Alicia makes dolls. She sells them for £5.95
How much does she get for 43 dolls?

Dividing decimals without a calculator

You need to set these out like you do whole number questions.
Keep your numbers in columns.

Example   Work out 8.75 ÷ 5

$$\begin{array}{r} 1. \\ 5\overline{)8.^375} \end{array}$$

First do **8 ÷ 5**. This is **1** with **3** left over.
Put the **1** above the **8** and carry the **3**.
Put the decimal point in.

$$\begin{array}{r} 1.\;7 \\ 5\overline{)8.^37^25} \end{array}$$

Now do **37 ÷ 5**
5 × 7 = 35. So 37 ÷ 5 is **7** with **2** left over.
Put the **7** above the **³7** and carry the **2**.

$$\begin{array}{r} 1.\;7\;5 \\ 5\overline{)8.^37^25} \end{array}$$

Now do **25 ÷ 5**
This is **5**. Put the **5** above the **²5**.

So   8.75 ÷ 5 = 1.75

Exercise 7:8

Work these out.

1 6.25 ÷ 5   **4** 4.74 ÷ 3   **7** 14.25 ÷ 3

2 8.25 ÷ 5   **5** 5.76 ÷ 4   **8** 18.64 ÷ 4

3 8.76 ÷ 3   **6** 12.75 ÷ 5   **9** 63.24 ÷ 6

10 A packet of ice creams costs £3.75
The packet contains 5 ice creams.
What is the cost of one ice cream?

11 A packet of 7 pieces of fish costs £6.65
Find the cost of 1 piece of fish.

12 A box of computer discs costs £4.25
The box contains 5 discs.
Find the cost of 1 disc.

13 Pete pays for 6 tickets to the cinema.
He pays £20.70
How much does each ticket cost?

Take your pick

  You need worksheet 7.1, **Take your pick**.

3 Estimating

· ·

100 000 attend Festival

Traffic was in chaos yesterday as 100 000 people arrived at Festival Park for an open air concert. Organisers were amazed at the number of people who came.

The attendance is given as 100 000. This does not mean that exactly 100 000 people came. The number has been rounded to give a sensible estimate. It would be silly for the paper to say 106 896 people came to the concert!

Rounding

Examples

1 Round 6.7 to the nearest whole number.

6.7 is nearer to 7 than to 6. It is rounded to 7.

2 Round 74 to the nearest 10.

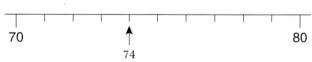

74 is nearer to 70 than to 80. It is rounded to 70.

3 Round 650 to the nearest 100.

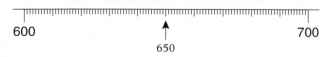

650 is halfway between 600 and 700. It is rounded to 700.

Exercise 7:9

1 Round these numbers to the nearest whole number.

a 5.7	**d** 9.4	**g** 13.7	**j** 34.7	**m** 124.6
b 3.6	**e** 4.5	**h** 23.6	**k** 46.5	**n** 136.2
c 7.1	**f** 9.8	**i** 24.2	**l** 58.5	**o** 152.8

2 Round these numbers to the nearest 10.

a 16	**d** 54	**g** 98	**j** 63	● **m** 134
b 34	**e** 67	**h** 14	**k** 36	● **n** 258
c 87	**f** 35	**i** 56	**l** 55	● **o** 312

3 Round these numbers to the nearest 100.

a 267	**d** 452	**g** 989	**j** 356	● **m** 2541
b 356	**e** 357	**h** 350	**k** 239	● **n** 3482
c 567	**f** 782	**i** 451	**l** 450	● **o** 5216

You often use your calculator to do questions.
When you do you should check that your answers are sensible.
It is very easy to hit the wrong key.

Example   Work out 5.4×179

Answer: 966.6

Estimate: 5.4 is about 5
            179 is about 200
            5.4×179 is about $5 \times 200 = 1000$
1000 is near to 966.6 so the answer is probably right.

Exercise 7:10

1 Copy these.
Fill in the missing numbers.
  a $5.7 \times 312 = \ldots$
      Estimate: 5.7×312 is about $\ldots \times \ldots = \ldots$
  b $6.9 \times 411 = \ldots$
      Estimate: 6.9×411 is about $\ldots \times \ldots = \ldots$

2 Work these out.
Write down the answer and an estimate for each one as in question **1**.
You need to decide how to estimate the numbers.
  a 4.6×218 **c** 3.8×67 **e** 58×394 **g** 499×303
  b 2.6×24 **d** 43×234 **f** 28×450 **h** 592×312

In the rest of this exercise:
  a work out the answers using a calculator,
  b write down an estimate to show that the answer is about right.

3 Chart CDs cost £13.49
Find the cost of 6 chart CDs.

4 A chair costs £49.85
Find the cost of 24 chairs.

5 There are 396 sweets in a jar.
How many sweets are there in 22 jars?

Prices in shops usually have 2 dp.
e.g.   £4.67   £40.36   £309.99
When you do questions with your calculator you sometimes get lots of
numbers after the decimal point.
When this happens you need to round your answers.

Example   A packet of 6 pizzas costs £3.25
Find the cost of 1 pizza.

1 pizza costs £3.25 ÷ 6

This is the answer on the calculator:

0.5416666

Draw a line after 2 dp. The next number is a 1.
Ignore all the rest of the numbers.

0.54|16666

0.541 is between 0.54 and 0.55

0.54 ↑
    0.541

0.541 is nearer to 0.54 than 0.55
So the answer is £0.54

Exercise 7:11

1   Round these calculator displays to 2 dp.

   a 2.1325728   **d** 3.5381449

   b 1.7836931   **e** 12.3378776

   c 5.4767911   **f** 24.7748919

2   Work these out.
Round your answers to 2 dp.

   a £4.58 ÷ 7   **c** £12.35 ÷ 6   **e** £45.37 ÷ 8
   b £9.93 ÷ 8   **d** £23.34 ÷ 9   **f** £565.71 ÷ 4

Example A box of choc ices costs £3.85
There are 8 choc ices in the box.
What is the cost of 1 choc ice?

1 choc ice costs £3.85 ÷ 8

3 **.** **8** **5** **÷** **8** **=** 0.48125

You need to round the display.
Answer: £0.48 or 48 p

Estimate: £3.85 is about £4
£3.85 ÷ 8 is about £4 ÷ 8 = £0.50
£0.50 is close to £0.48 so the answer is probably right.

In the rest of this exercise:
a work out the answers using a calculator,
b write down an estimate to show that the answer is about right.

3 A bag of chicken pieces costs £4.89
The bag contains 8 pieces of chicken.
What is the cost of 1 piece?

4 Jenny buys a bag of fun size Mars bars.
She pays £2.19
There are 18 Mars bars in the bag.
How much is 1 Mars bar?

5 Phil buys a bag of cooking apples.
He pays £1.48
There are 12 apples in the bag.
How much is 1 apple?

1 Nadia enters a discus competition.
These are the lengths of her throws:

7.68 m   6.5 m   7.45 m   8.34 cm   8.387 m

Put her throws in order of size. Start with the biggest.

2 These are the times in seconds for the
women's 400 m final in Atlanta.

G. Breuer 50.71   F. Ogunkoya 49.10
P. Davis 49.28   M-J Perec 48.25
C Freeman 48.63   S. Richards 50.45
J. Miles 49.55   F. Yusuf 49.77

a What is the winning time?
b List the times in order. Start with the smallest.
c How many seconds are there between first and second?
d How many seconds are there between first and last?

3 Leroy is going to the fair. It costs £3.25 to get in.
Every ride is then only 20 p.
What will it cost Leroy if he has 8 rides?

4 This is a price list for sweatshirts:

	small	medium	large	extra large
plain	£7.25	£7.75	£7.95	£8.25
pattern	£7.75	£8.25	£8.45	£8.75

a How much more does it cost for a large
   patterned sweatshirt than for a large plain?
b Glenda buys 3 medium patterned sweatshirts.
   How much does she pay?
c Gill buys 2 small plain and 2 small patterned
   sweatshirts. How much does she pay?

5 Lil makes purses to sell.
She sells the purses for £2.65 each.
How much does she get if she sells 27 purses?

6 Stamps cost 26 p. Jill has £6.
   a Work out how many stamps she can buy. Use a calculator.
   b Write down an estimate to show that the answer is about right.

1 Juliette enters a diving competition.
There are 5 judges who give her a mark.
These are the marks:

  7.5   7.7   7.9   7.3   8.1

The highest and the lowest marks are ignored.
a Which marks are ignored?
The other 3 marks are added to give a total.
b What is the total for Juliette's dive?
This score is then multiplied by the dive rating
to give the score.
The dive rating for Juliette's dive is 3.
c What is Juliette's score?
d Rachel is in the same competition. Her dive rating is 3.
These are her marks:

  6.2   7.0   7.1   6.4   6.8

What is Rachel's score?

2 Jason has drawn this triangle.
He says that the perimeter of the
triangle is about 15 cm.
a How can you tell that he is
wrong without measuring the
other two sides?
b Measure the lengths of the other
two sides.
Write your answers in centimetres.
c Add up the lengths of the three
sides to find the perimeter of the
triangle.

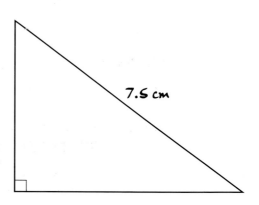

7.5 cm

3 The numbers in ◯ and ◯ add together to give ▢ like this:

(3.2)——|8.3|——(5.1)

Fill in the missing numbers:

a (2.8)——| |——(3.3)

b ◯——|6.4|——(2.7)

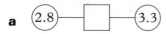

- **Putting decimals in order of size**

  Examples　**1**　3.12 is smaller than 4.34
  because 3 is smaller than 4
  2　3.463 is smaller than 3.58
  because 4 is smaller than 5
  3　3.441 is smaller than 3.45
  because the 4 is smaller than the 5

- **Adding and subtracting decimals**

  Examples

  $$
  \begin{array}{r}
  63.36 \\
  +\ 24.90 \\
  \hline
  88.26 \\
  \end{array}
  $$

  $$
  \begin{array}{r}
  5\overset{3}{\cancel{4}}.\overset{1}{2}5 \\
  -\ 33.44 \\
  \hline
  20.81 \\
  \end{array}
  $$

- **Multiplying decimals**

  Examples

  $$
  \begin{array}{r}
  4.73 \\
  \times\quad 5 \\
  \hline
  23.65 \\
  \end{array}
  $$

  $$
  \begin{array}{r}
  3.76 \\
  \times\quad 24 \\
  \hline
  75.20 \\
  15.04 \\
  \hline
  90.24 \\
  \end{array}
  $$

- **Dividing decimals**

  Example

  $$
  \begin{array}{r}
  1.\ 7\ 5 \\
  5\overline{)8.^37^25} \\
  \end{array}
  $$

- **Estimating**　Work out 5.4×179

  Example

  | 5 | . | 4 | × | 1 | 7 | 9 | = |

  Answer: 966.6
  Estimate:　5.4 is about 5
  　　　　　179 is about 200
  　　　　　5.4×179 is about $5 \times 200 = 1000$
  1000 is near to 966.6 so the answer is probably right.

- **Using a calculator**　A box of choc ices costs £3.85
  　　　　　　　　　There are 8 choc ices in the box.

  Example　What is the cost of 1 choc ice?
  1 choc ice costs £3.85 ÷ 8

  | 3 | . | 8 | 5 | ÷ | 8 | = |　　0.48125

  You need to round the display.
  Answer: £0.48

1 These are the times in seconds for the men's 400 m final in Atlanta.

R. Black 44.41   D. Kamoga 44.53
D. Clarke 44.99   R. Martin 44.83
A Harrison 44.62   I. Thomas 44.70
M. Johnson 43.49

a What is the winning time?
b List the times in order. Start with the smallest.
c How many seconds are there between first and second?
d How many seconds are there between first and last?

2 Write these decimals in order of size.
Start with the smallest.
a 3.67   4.29   4.27   **b** 1.64   2.23   2.201

3 Work these out:
a £34.70 + £21.20   **e** £2.34 × 3
b £4.80 + £1.40   **f** £12.45 × 15
c £5.27 − £4.15   **g** £4.59 ÷ 3
d £23.47 − £12.73   **h** £5.75 ÷ 5

4 A bag of fish pieces costs £3.29
The bag contains 6 pieces of fish.
a What is the cost of 1 piece?
Use a calculator.
b Write down an estimate to show
that the answer is about right.

5 Year 9 classes in Stanthorne High collect money for charity.
They have collected £978.86
Copy this headline from the local paper.
Fill in the missing number.

Local school raises nearly £⬛⬛⬛ for charity

8 Algebra

1	1	1	1
1	3	5	7
1	5	13	25
1	7	25	?

What is the missing number?

1   Warp factor one!

Nathan and his friends are at Alton Towers.
They are queuing for the Nemesis ride.
You have to go on the ride in groups of four.
There are 17 people in Nathan's group.
Can they all get on the ride together?

Exercise 8:1

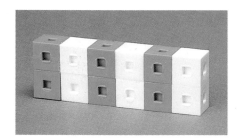

You will need some cubes and a worksheet for this exercise.
You are going to make rectangles with the cubes.
Keep your answers to this exercise. You will need them in Exercise 8:3.

1   **a**   Get 12 cubes. Put them in a single line on your desk.

    b   Put the cubes on the worksheet so that they are in two rows.
How many columns are there?

    c   Now put the cubes into three rows.
How many columns are there?

    d   Put the cubes into four rows.
How many columns are there now?

    e   Copy this table. Fill it in up to row 4.

12	Number of rows	Number of columns
	1	
	2	
	3	
	4	
	5	
	6	

f Put your cubes into five rows.
   They will not make a rectangle. Shade this space in your table.
g Now try six rows. Fill in row 6 of your table.
h There is one other way to make the cubes into a rectangle.
   Try to find it.
   Fill in the last row of the table.

2 Get 8 cubes. Put them in a single row on your desk.
   a Copy this table.

8	Number of rows	Number of columns
	1	
	2	
	3	
	4	
	5	
	6	
	7	
	8	

b Put the cubes into two rows on the worksheet.
   Do they make a rectangle?
   Fill in the number of columns in the table.
c Now try to put the cubes into three rows. Do they make a rectangle?
   You should find there is a gap in one column.
   Shade in row 3 of your table.
d Now try four rows.
   If it works, fill in the table. If it does not work, shade in row 4.
e Fill in the rest of the table.

3 Get 15 cubes. Put them in a single row to start with.
   a Copy this table.

15	Number of rows	Number of columns
	1	

b Try to make other rectangles with the cubes.
   Fill in the table **only** when they make a rectangle.

Exercise 8:2

You are now going to investigate the rectangle problem with different numbers of cubes.

You should:

(1) Get the right number of cubes.

(2) Put them in a single row on your desk.

(3) Draw a table in your book like this one:

	Number of rows	Number of columns
	1	

(4) Put the number of cubes in the corner box.
    Fill in row 1 first. This will always work.

(5) Try to put the cubes into two rows, then three, then four and so on.
    Only fill the answers in your table when the cubes make a rectangle.
    Make sure that you always try each number of rows.

Keep your answers to this exercise.
You will need them for the next exercise.
Here are the numbers that you should try.

Set A:	20	18	21
Set B:	9	16	25
Set C:	11	17	13

When you have finished all the numbers, answer these questions:

1 For Set A: Can you always make more than two rectangles?

2 For Set B: There is something special about these numbers.
Can you say what?

3 Write down what you found out about Set C.
Do you know the special name for these numbers?

| Factor | A number that divides exactly into another number is called a **factor**. |

Examples

$14 = 1 \times 14$
$14 = 2 \times 7$

1, 14, 2 and **7** are **factors** of 14

$16 = 1 \times 16$
$16 = 2 \times 8$
$16 = 4 \times 4$

1, 2, 4, 8 and **16** are **factors** of 16

Exercise 8:3

1 Look back at your answers to Exercise 8 : 1 question **1**.
Write down all the factors of 12.

2 Look back at your answers to Exercise 8 : 1 question **2**.
Write down all the factors of 8.

3 Look back at your answers to Exercise 8 : 1 question **3**.
Write down all the factors of 15.

Use your answers to Exercise 8 : 2 to help you with these questions.

4 Write down all the factors of 20.

5 Write down all the factors of 18.

6 Write down all the factors of 21.

7 Write down all the factors of 25.

8 Write down all the factors of 9.

Exercise 8:4

You can use cubes to help you with this exercise. Try to use them as little as possible. You might find a calculator helpful.

1 a Copy this table.

24	Number of rows	Number of columns
	1	24
	2	
	3	
	4	

b Fill in the table.
c Write down all the factors of 24.

2 a Copy this table.

6	Number of rows	Number of columns
	1	6
	2	

b Fill in the table.
c Write down all the factors of 6.

3 a Copy this table.

30	Number of rows	Number of columns
	1	
	2	
	3	
	5	

b Fill in the table.
c Write down all the factors of 30.

Prime numbers Prime numbers have only two factors, themselves and 1.

Example

$13 = 1 \times 13$
No other two whole numbers multiply to give 13.
13 is a prime number.

Some other prime numbers are 3, 5 and 7.

1 is **not** a prime number.

Exercise 8:5

1 Look back at your answers for Exercise 8:2 Set C.
Write down three prime numbers.
These are the numbers which only made one rectangle.

2 a Copy this grid of numbers:

1	2	3	4	5
6	7	8	9	10
11	12	13	14	15
16	17	18	19	20
21	22	23	24	25

b Cross out the 1. It is not a prime number.
c Put a circle around the 2.
Divide all the other numbers in the grid by 2.
If a number divides exactly by 2 cross it out.
d Now circle the 3.
Divide all the numbers that are left by 3.
Cross out the ones that divide exactly.
e Circle the 5 and do the same.
f Circle all the numbers that are left.
These are all prime numbers.

3 Which of these numbers are prime numbers?
a 17   **d** 19   **g** 31
b 11   **e** 26   **h** 53
c 15   **f** 30   **i** 57

You can always split a number up into prime numbers multiplied together.

$15 = 3 \times 5$
$27 = 3 \times 3 \times 3$
$36 = 2 \times 2 \times 3 \times 3$

This is called finding the **prime factors** of a number.
The best way to do this is using a factor tree.

Example   Use a factor tree to find the prime factors
of 40.

(1) Write down the number and put it in
a box:

(2) Think of any two factors that multiply
to make this number.
Put these underneath your box.
Connect them with lines.
These are the branches of the tree.

(3) Circle any prime numbers.

(4) Look at the numbers you have not
circled.
Split these into a factor pair.

(5) Continue until all the branches end
in circles.
The numbers in the circles are the
prime factors of the starting number.

(6) Write out the factors as a multiplication.
$40 = 2 \times 2 \times 2 \times 5$

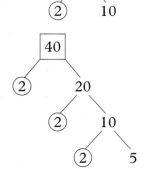

---

4   For each part
(1) Copy the factor tree.
(2) Finish off the tree.
(3) Write out the factors as a multiplication.

a   **b**   **c**   **d**

2   Dividing it up

It is 9R's turn to stack the chairs after assembly.
The chairs must be stacked in piles of 8.
If they were any higher then they might fall over.
There are 320 chairs to stack altogether.
Will the chairs fit into an exact number of stacks of 8?

Exercise 8:6

You will need a calculator for this exercise.

1 a On your calculator press **8** **+** **+** **=**
   b Now press **=** again.
   c Write down the number you get.
   d Press **=** again.
   e Write down the number you get.
   f Use your calculator to find the next 10 numbers in this pattern.
   g Keep pressing the **=** sign until you get above 300.
      Slow down!! Can you get to exactly 320?
   h Will the 320 chairs stack exactly into piles of 8?

Multiples	All the numbers you wrote down in question **1** are **multiples** of 8. This means that 8 will divide into them exactly.
Example	3 6 9 12 15 18 are multiples of 3
	25 is not a multiple of 3 because 3 does not go exactly into 25

---

2 a On your calculator press $\boxed{5}$ $\boxed{+}$ $\boxed{+}$ $\boxed{=}$
b Now press $\boxed{=}$ again.
c Write down the number you get.
d Press $\boxed{=}$ again.
e Write down the number you get.
f Use your calculator to find the next 10 numbers in this pattern.
g Copy and fill in:
These numbers are all multiples of … .

3 a On your calculator press $\boxed{7}$ $\boxed{+}$ $\boxed{+}$ $\boxed{=}$
b Write down the first 10 multiples of 7.

4 Use your calculator to work out the first 10 multiples of 9.

5 a Press $\boxed{2}$ $\boxed{+}$ $\boxed{+}$ $\boxed{=}$ on your calculator.
b Write down 2.
c Press $\boxed{=}$ and write down the number you get.
d Use your calculator to find the next 10 numbers in this pattern.
e What is the special name for these numbers?
Choose from **odd   even   prime   square**
f Copy and fill in:
Even numbers are all multiples of … .

6 Ned has 120 computer disks. He wants to pack them into boxes of 5.
a Press $\boxed{5}$ $\boxed{-}$ $\boxed{-}$ $\boxed{=}$ on your calculator.
b Every time you press $\boxed{=}$ the calculator will take 5 off the total.
This is like filling one box.
Press $\boxed{1}$ $\boxed{2}$ $\boxed{0}$ $\boxed{=}$
c Work out how many boxes Ned can fill.
d Will he have any disks left over?
Explain how you know.

7 Danielle has 75 p. She wants to buy some chocolate bars.
They cost 15 p each.
a Press $\boxed{15}$ $\boxed{-}$ $\boxed{-}$ $\boxed{=}$ on your calculator.
b Every time you press $\boxed{=}$ it is like buying one bar.
Press $\boxed{7}$ $\boxed{5}$ $\boxed{=}$
c Work out how many chocolate bars Danielle can buy.
d Will she have any money left?

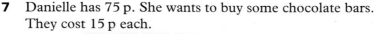

Exercise 8:7

▼ *Countdown!*

You will need a worksheet to play this game.

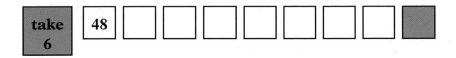

1 Set your calculator up to take away the number in the box.
If you want to take away 6 press **6** **−** **−** **=**

2 The aim is to count down to 0.
Colour in the box where you think you will get down to 0.

3 Put the start number into your calculator.

4 Press **=** and write the answer in the first empty box in the row.

5 Keep pressing **=** and writing down the answer until you get to 0.

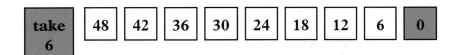

6 Scoring:

If you get to 0 on your coloured square score 5 points.
If you are one square out score 3 points.
If you are two squares out score 1 point.
Otherwise score 0 points.

There are five games on the sheet.

You can ask for a blank sheet from your teacher so that you can
make up some games of your own.

If you need to find a factor of a big number it is quicker to divide.

Examples **1** Find a factor of 453 other than 1.

Try 2: $453 \div 2 = 226.5$
2 doesn't go into 453 exactly. It is not a factor.

Try 3: $453 \div 3 = 151$
3 goes into 453 exactly. 3 is a factor of 453.

2 Find a factor of 391.

Try 2: $391 \div 2 = 195.5$
Try 3: $391 \div 3 = 130.333\,333\,33$
Try 4: $391 \div 4 = 97.75$

Try 16: $391 \div 16 = 24.4375$
Try 17: $391 \div 17 = 23$ So 17 is a factor. So is 23.

Exercise 8:8

1 **a** Work out $201 \div 2$.
  b Is 2 a factor of 201?
  c Work out $201 \div 3$.
  d Is 3 a factor of 201?
● **e** Look at your answer to **c**.
      Write down another factor of 201.

2 **a** Work out $203 \div 2$.
  b Is 2 a factor of 203?
  c Work out $203 \div 3$.
  d Is 3 a factor of 203?
  e Work out $203 \div 5$.
  f Is 5 a factor of 203?
  g Work out $203 \div 7$.
  h Is 7 a factor of 203?

3 Find one factor other than 1 of each of these numbers:
  a 275   **b** 161   **c** 136

Sometimes it is possible to spot factors very easily.

Exercise 8:9

1 **a** Copy this table. Fill it in.

Number	2 × number	5 × number	10 × number
1	2	5	10
2	4	10	20
3	6	15	30
4	8	20	40
5			
6			
7			
8			
9			
10			

  b Look at the red numbers in the 2 × column.
     All the numbers end in 0, 2, 4, 6 or 8.
     What is the name for these numbers?
  c Look at the green numbers in the 5 × column.
     Copy and fill in:
        The numbers in the 5 × column always end in … or … .
  d Look at the blue numbers in the 10 × column.
     Copy and fill in:
        The numbers in the 10 × column always end in … .

2 Copy these sentences. Fill them in.
  a If 2 is a factor of a number, it will end in ....
  b If 5 is a factor of a number, it will end in ....
  c If 10 is a factor of a number, it will end in ....

3 **a** Which of these numbers have 2 as a factor?
         156   418   457   638
  b Which of these numbers have 5 as a factor?
         155   327   125   140   259
  c Which of these numbers have 10 as a factor?
         130   147   6970   125   367

3 Square and triangle numbers

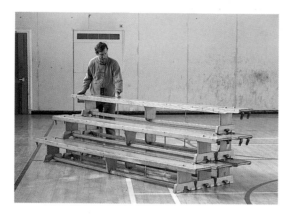

Year 9 are having a year photograph taken.
Len Scap the photographer has stacked some PE benches for them to stand on.
If he has to add another row at the back, how many benches would there be altogether?

Exercise 8:10

You will need some cubes for this exercise.

1 a Build this shape out of cubes.
  b How many cubes did you use?

2 a Build this shape out of cubes.
  b How many cubes did you use?

3 a Build the next model in this pattern.
  b How many cubes did you use?

4 **a** Copy this table.

Model number	Number of cubes
1	3
2	6
3	
4	
5	
6	

b Fill in the table.
You can build the models to help you.

The numbers in the second column of the table have a special name.
They are called **triangle numbers**.

Triangle numbers	1 3 6 10 15 21 ... are called **triangle numbers**.

You can build a triangular shape out of these numbers of cubes.

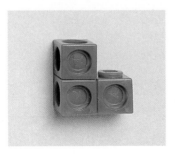

5 **a** Copy this:

```
1                    = 1
1 + 2                = 3
1 + 2 + 3            = 6
1 + 2 + 3 + 4        = 10
1 + 2 + 3 + 4 + 5    = 15
```

b Add another four rows.

Two consecutive triangles will fit together to make a square.

Exercise 8:11

1 a Make these models.
Use different coloured cubes.

b Fit the two models together.
They should make a square.

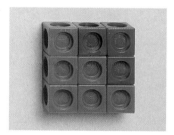

c How many cubes have you used altogether?

2 a Keep your bigger triangle.
Make the next size from a different colour.
b Fit the two models together.
They should make a square.
c How many cubes have you used altogether?

● **3 a** Make the next size of square out of two triangles.
b How many cubes have you used altogether?

● **4** Make the next size of square.
How many cubes have you used?

Square numbers	1 4 9 16 ...
	are called **square numbers**.

You can build a square shape out of these numbers of cubes.

There are lots of patterns you can make with square numbers.

Exercise 8:12

1 a Copy this pattern of numbers:

1	= 1
1 + 2 + 1	= 4
1 + 2 + 3 + 2 + 1	= 9
1 + 2 + 3 + 4 + 3 + 2 + 1	= 16

   b Add another three rows to the pattern.
   c Copy these diagrams.

   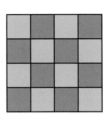

   d Draw the next diagram in the pattern.
   e How many squares would there be in the sixth diagram?

2 a Copy this pattern of numbers:

$$1 \times 1 = 1$$
$$2 \times 2 = 4$$
$$3 \times 3 = 9$$
$$4 \times 4 = 16$$

   b Add the next five rows to this pattern.

3 a Copy this pattern of numbers:

1	= 1
1 + 3	= 4
1 + 3 + 5	= 9
1 + 3 + 5 + 7	= 16

   b Add another three rows to the pattern.

c Copy these diagrams.

d Draw the next diagram in the pattern.

4 Look at these diagrams:

a Draw the next three diagrams in this pattern.
Use the same colours as in the book.

b Can you find any patterns in these numbers?

1	= 1
4	= 4
1 + 8	= 9
4 + 12	= 16
... + ...	= ...

Squaring a number	Squaring a number means multiplying it by itself. 3 squared is 9. This is because $3 \times 3 = 9$. You can write 3 squared as 3^2.

Example

Find the values of **a** 5^2 **b** 7^2 **c** 25^2

a $5^2 = 5 \times 5 = 25$

b $7^2 = 7 \times 7 = 49$

c $25^2 = 25 \times 25 = 625$

Exercise 8:13

1 Work these out.

a 4^2 **c** 9^2 **e** 20^2
b 8^2 **d** 12^2 **f** 30^2

You can use the x^2 button on your calculator to square numbers.

Example   Work out 3.4^2

Key in   **3** **.** **4** **x^2** **=**

The answer is 11.56

2   Work these out. Use the x^2 button on your calculator.

   a 40^2 **c** 900^2 **e** 0.5^2
   b 7.3^2 **d** 1.56^2 **f** 0.2^2

You can use the same sort of method if you want to multiply a number by itself lots of times.
You can write $2 \times 2 \times 2 \times 2 \times 2$ as 2^5.
This means that $2^5 = 32$.

Power   In 2^5, the 5 is called the **power**.
The power tells you how many 2s to multiply together.

Example   Work out 4^3.
$4^3 = 4 \times 4 \times 4 = 64$

3   Write these numbers using a power.

   a 3×3 **d** $8 \times 8 \times 8 \times 8$
   b $5 \times 5 \times 5$ **e** $9 \times 9 \times 9 \times 9 \times 9 \times 9$
   c $6 \times 6 \times 6 \times 6 \times 6$ **f** $4 \times 4 \times 4 \times 4 \times 4 \times 4 \times 4$

You can use the y^x button on your calculator to work out numbers.

Example   Work out 5^6.

Key in   **5** **y^x** **6** **=**

The answer is 15625.

4   Work these out. Use the y^x button on your calculator.

   a 5^4 **d** 6^4 **g** 8^3 **j** 15^3
   b 3^6 **e** 4^7 **h** 6^5 **k** 4.1^3
   c 7^3 **f** 2^8 **i** 21^2 **l** 3.7^5

1 Look at this rectangle of 12 cubes.

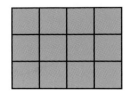

    a What is the width of the rectangle?
    b Write down one factor of 12.
    c Use the diagram to write down another factor of 12.
    d Draw a different rectangle using 12 cubes.
    e Write down two more factors of 12.

2 Look at this list of numbers:
    12   8   9   13   24   6
    Only **one** of these numbers is a prime number. Which number is prime?

3 Daniel is planting 18 bulbs in his garden.
    He can plant them in three rows of 6, like this:

    • • • • • •

    • • • • • •

    • • • • • •

    a Draw a diagram to show how he could plant the bulbs in
       two rows of 9.
    b Draw a diagram to show a **different** way he can plant these 18 bulbs.
       There must be the **same** number of bulbs in each row.
    c Copy this table. Fill it in.

Number of rows	Number of bulbs in each row
1	
2	9
3	6
6	
9	
18	

    d Sonya thinks she can plant 18 bulbs in rows of 5 with the same
       number of bulbs in each row. Explain why she is wrong. You can
       draw a diagram.

4 William draws a triangle using dots:

Claire draws a bigger triangle:

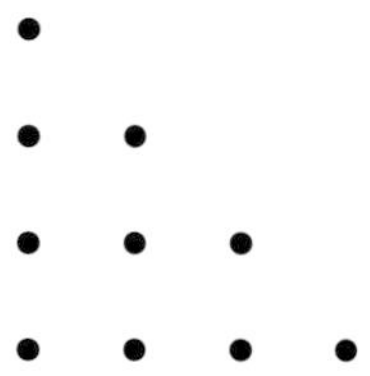

  a How many dots did William use?
  b How many dots did Claire use?
  c Draw the next triangle in the pattern.
  d How many dots did you use?
  e Write down the number of dots in the next **three** triangles in the pattern.

5 A block of flats has post boxes in its entrance hall.
They have the flat numbers on them.

1	6	11	16	
2	7	12	17	
3	8	13	18	
4	9	14	19	
5	10	15	20	

  a What would be the next number on the top row?
  b What would be the next number on the middle row?
  c Write down the next **three** numbers on the bottom row.
  d Copy and fill in:

   The numbers on the bottom row are all .... of five.

  Choose from **factors, multiples, primes, products**

1 Paula has made some sweets to sell at the school fair.
She has 300 sweets. She is packing them into bags of ten.

Paula wants to know how many bags she will need.
She works out $300 \times 10 = 3000$.
a Why **must** this answer be wrong?
b What should Paula have done?
c How many bags will Paula need?

2 Dave is packing cakes into boxes of five.
He has 34 cakes.
Will he be able to fill an exact number of boxes?
Explain your answer.

3 Here are some cards with numbers on them.

120	17	435	64	29
690	16	1	15	99

a Which of the cards show **multiples** of 10?
b Which of the cards show **multiples** of 5?
c Which of the cards show **square** numbers?
d Which of the cards show **prime** numbers?

4 This rectangle has an area of 24 cm².

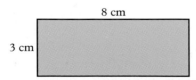

8 cm

3 cm

a Draw another rectangle that has an area of 24 cm².
b Draw a **different** rectangle with an area of 24 cm².
c Write down all the factors of 24.
d Draw as many rectangles as you can that have an area of 30 cm².
e Write down all the factors of 30.

- **Factor**

  A number that divides exactly into another number is called a **factor**.

  Example $14 = 1 \times 14$
  $14 = 2 \times 7$

  1, 2, 7 and **14** are **factors** of 14

- **Prime numbers**

  Prime numbers have only two factors, themselves and 1.

  Example $13 = 1 \times 13$
  No other two whole numbers multiply to give 13.
  13 is a prime number.

  Some other prime numbers are 3, 5 and 7.

  1 is **not** a prime number.

- **Multiples**

  All these numbers are **multiples** of 3.
  This means that 3 will divide into them exactly.

  3   6   9   12   15   18

  25 is not a multiple of 3 because 3 does not go exactly into 25.

- **Triangle numbers**

  1   3   6   10   15   21
  are called **triangle numbers**.

  You can draw triangular shapes out of these numbers of dots:

- **Square numbers**

  1   4   9   16   ...
  are called **square numbers**.

  You can draw square shapes with these numbers of dots:

---

1 Get 16 cubes or counters. Put them in a single row.
 a Copy this table.

16	Number of rows	Number of columns
	1	16
	2	

 b Put the cubes into two rows.
   Do they make a rectangle? Fill in the number of columns in the table.
 c Now try to put the cubes into three rows. Do they make a rectangle?
   If it works, fill in the table.
 d Now try four rows.
   If it works, fill in the table.
 e Make as many different rectangles as you can.
   Fill in the sizes in your table.
 f Write down all the factors of 16.

2 **a** Work out 396 ÷ 2.
 b Is 2 a factor of 396?
 c Work out 396 ÷ 3.
 d Is 3 a factor of 396?
 e Work out 396 ÷ 5.
 f Is 5 a factor of 396?
 g Find as many other factors of 396 as you can.

3 Look at this list of numbers:
   3   8   9   16   19   21   30   31
 a Write down the two **square** numbers.
 b Write down the three **prime** numbers.
 c Write down the **multiples** of 4.

4 Here are the patterns for three triangle numbers.

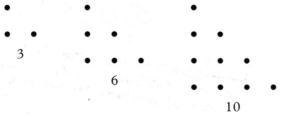

Draw the next three triangle patterns.
Write the triangle numbers underneath.

9 Time

QUESTIONS

EXTENSION

SUMMARY

TEST YOURSELF

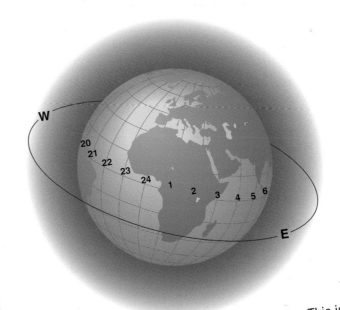

The basic unit of time measurement is the day. This is one rotation of the Earth on its axis. Dividing the day into hours and minutes and seconds is only for our convenience.

The second is defined in terms of the resonance vibration of the caesium-133 atom, as the interval occupied by 9 192 631 770 cycles.

1 Planning time

When you cook you have to plan time.
This cake had to be cooked in time for a birthday party.
It had to be cooked for the correct length of time.
The cake would not be nice if it was underdone or burnt.

Sally's birthday is 29th April. She is looking at this calendar to see what day of the week her birthday will be.

FEBRUARY						
S	M	T	W	T	F	S
...	...	1	2	3	4	5
6	7	8	9	10	11	12
13	14	15	16	17	18	19
20	21	22	23	24	25	26
27	28	29	...	...	...	...

MARCH						
S	M	T	W	T	F	S
...	...	...	1	2	3	4
5	6	7	8	9	10	11
12	13	14	15	16	17	18
19	20	21	22	23	24	25
26	27	28	29	30	31	...

APRIL						
S	M	T	W	T	F	S
...	...	...	...	...	...	1
2	3	4	5	6	7	8
9	10	11	12	13	14	15
16	17	18	19	20	21	22
23	24	25	26	27	28	**29**
30	...	...	...	...	...	...

Exercise 9:1

1 Look at Sally's calendar.
   a The letters S M T W T F S are at the top of each month.
     What do the letters stand for?
   b Sally's birthday is 29th April.
     It is marked in red.
     Use Sally's calendar to find what day of the week her birthday will be.

2 Write down the days of the week for each of these dates.
   Use Sally's calendar.
   a 3rd February   b 19th March   c 1st April   d 28th February

1 ordinary year = 365 days   1 leap year = 366 days

Leap years are …, 2000, 2004, 2008, 2012, 2016, …

3 Does Sally's calendar show part of a leap year?
Explain how you found the answer.

4 The numbers of the leap years make a pattern:

    a Write down the rule for the pattern.
    b Give the next two numbers in the pattern.

5 Only three months are shown here but Sally's calendar is for a whole year.
    a What day of the week is the 31st January on Sally's calendar?
    b What day of the week is 1st May shown on Sally's calendar?

6   **a** How many days are there in April?
    b Write down all the months in a year with the same number of days as April.

7   **a** Look at the dates of the Tuesdays in February.
       They form this number pattern:

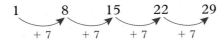

       Explain why the rule is +7.
    b Write down the number pattern for the Fridays in March.
       Do they follow the same rule?
    c Write down the number pattern for the Sundays in April.
       Do they follow the rule?

8 Here is a number pattern for the Saturdays in April.
They start with Sally's birthday.

    a What is the rule for this number pattern?
    b Sally's mother ordered a birthday cake for Sally exactly four weeks before
       her birthday.
       Write down the date that she ordered the cake.

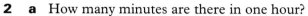

Exercise 9:2

1 **a** How many seconds are there in one minute?
    b How many seconds are there in two minutes?
    c The sand takes three minutes to run through
       this egg timer.
       How many seconds does the sand take to run
       through the egg timer?
    d How many seconds are there in 10 minutes?

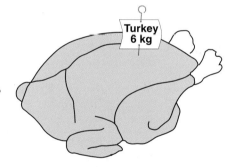

2 **a** How many minutes are there in one hour?
    b How many minutes are there in two hours?
    c Sally's birthday party is to last from 7 pm until 11 pm.
       How many hours will the party last?
    d How many minutes will Sally's party last?

3 Sally's brother Peter has a 180 minute video tape to video the party.
    How many hours will the video tape last?

4 Sally is roasting this turkey for the party.
    The instructions say 30 minutes per
    kilogram plus 30 minutes.
    a How long does the turkey take to cook
       in minutes?
    b How long is this in hours and minutes?

5 Sally's mother is cooking a piece of ham for the party.
    The ham takes 40 minutes per kilogram plus 30 minutes.
    The ham weighs 3 kg.
    a How long does it take to cook in minutes?
    b How long is this in hours and minutes?

6 **a** How many hours are there in one day?
    b How many hours are there in two days?
    c How many hours are there in one week?
    d Sally's father says there are 72 hours to go until Sally's birthday.
       How many days is it until Sally's birthday?

7 The birthday cake was ordered four weeks before Sally's birthday.
    How many days before her birthday was this?

Times after midnight and in the morning are am.
Times in the afternoon and evening are pm.

It is nearly time for Sally's party.

Peter says the time is 6.40 pm.

Sally says the time is twenty to
seven in the evening.

They are both right.

Exercise 9:3

1 Write down the times shown on these clocks.
Use am and pm (Peter's method).

a

afternoon

c

morning

e

evening

b

evening

d

morning

f
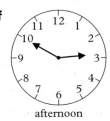
afternoon

2 Write down the times shown on the clocks in question **1**.
Use Sally's method.

3 Write these times as am or pm times.
   a Five past four in the afternoon.   **c** A quarter past five in the morning.
   b Ten past nine in the morning.   **d** Half past one in the afternoon.

* * *

Example   Write eight minutes to five in the afternoon as an am or pm time.

The time is between 4 pm and 5 pm.

You need to work out the
number of minutes after 4 pm.
   $60 - 8 = 52$

The time is 4.52 pm.

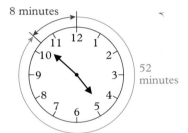

* * *

4 Write these times as am or pm times.
   a Ten to three in the afternoon.   **c** Five to nine in the evening.
   b Twenty to two in the afternoon.   **d** Six minutes to ten in the morning.

5 These things happened during Sally's birthday.
   Copy the table.
   Fill it in.

	What happened	am/pm time	Time in words
a	Woke up	...	twenty to eight
b	Got up	7.55 am	five to eight
c	Had breakfast	8.25 am	...
d	Opened presents	...	quarter to nine
e	Telephoned friend	9.40 am	...
f	Had lunch	12.55 pm	...
g	Went to shop	...	twenty past two
h	Sorted tapes	...	twenty-five to four
i	Got ready for the party	5.45 pm	...
j	First guest arrived	6.52 pm	...
k	Ate party food	...	half past eight
l	Last guest went home	11.07 pm	...

Example

Peter is cooking chicken portions.
The chicken portions go in the oven at 5.40 pm.
They will take **35** minutes.
When will they be done?

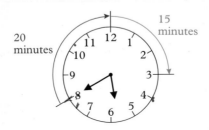

At 5.40 pm it is **20** minutes to the next hour.
$35 - 20 = 15$
The chicken will be done at 6.15 pm.

Exercise 9:4

1 This pizza is put in the oven at
6.05 pm.
When does the pizza need to be
taken out of the oven?

2 Work out the times when each of these will be cooked.

	Food to be cooked	Time put in oven	Time taken to cook
a	Small cakes	3.10 pm	15 minutes
b	Sponge cake	10.18 am	25 minutes
c	Spareribs	7.09 pm	45 minutes
d	Fruit pie	6.20 pm	45 minutes
e	Shepherds' pie	5.45 pm	35 minutes
f	Apple crumble	12.37 pm	40 minutes
g	Chocolate cake	10.52 am	50 minutes

3 This steak pie is put in a hot oven at 5.42 pm.
When will it be done?

Steak pie

Brush with beaten egg or milk. Bake in a pre-heated oven at 200 °C, (400 °F) for 45 minutes

4 Sally has to cook this vegetable curry. The time is 5.55 pm.
   a When will the curry be cooked if Sally uses a microwave oven?
   b Sally decides to use the ordinary oven. She knows the oven will take 10 minutes to warm up. When will the curry be cooked?

Vegetable curry

Place in a pre-heated oven at 200 °C, 400 °F for 35 minutes. To microwave: cook on full power for 7 minutes.

Vegetable curry

Example   Sally's mother is cooking a chicken. It takes 1 hour 45 minutes to cook. She puts the chicken in the oven at 9.50 am. When will it be cooked?

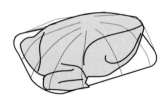

Sort the 45 minutes out first.   At 9.50 am it is **10 minutes** to the next hour.
45 − 10 = 35

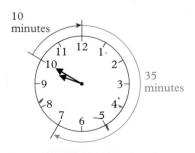

10 minutes

35 minutes

*If the chicken took **45 minutes** it would be ready at 10.35 am.*

   10.35 + 1 hour = 11.35 am

The chicken will be cooked at 11.35 am.

---

5 Sally is cooking a fruit cake.
She puts the fruit cake in the oven at 9.40 am.
The cake takes 3 hours 30 minutes.
When will the cake be done?

Exercise 9:5

The blue cards show:
  something to be cooked;
  the time it goes in the oven;
  how long it needs.

Match each blue card to a finishing time.

1
> *Leg of lamb*
> Into oven 3.50 pm
> 2 hours 15 minutes

5
> *Chicken*
> Into oven 11.45 am
> 1 hour 20 minutes

Finishing times

6.05 pm

12.15 pm

7.10 pm

2
> *Lasagne*
> Into oven 5.45 pm
> 1 hour 25 minutes

6
> *Meringues*
> Into oven 10.55 am
> 3 hours 15 minutes

12.40 pm

7.15 pm

1.05 pm

3
> *Beef stew*
> Into oven 9.55 am
> 2 hours 45 minutes

7
> *Jacket potatoes*
> Into oven 10.25 am
> 1 hour 50 minutes

2.10 pm

6.15 pm

4
> *Orange cake*
> Into oven 4.35 pm
> 1 hour 40 minutes

8
> *Egg and bacon flan*
> Into oven 5.40 pm
> 1 hour 35 minutes

2 Timetables

Alan lives in London. His cousin Sally lives in Matlock near Derby. Alan is going to Sally's party. He is travelling by train to Derby. He is looking at the train timetable.

Exercise 9:6

1 Look at this train timetable.
  a What time does the first train leave London?
  b What time does this train arrive in Derby?

2 What time does the 12 30 from London arrive in Derby?

3 Alan arrives at the station in London at 09 45. What is the earliest time he can catch a train to Derby?

London depart	Derby arrive
0730	0936
0830	1036
0930	1141
1030	1235
1130	1336
1230	1440
1330	1535
1430	1636
1530	1736
1630	1848
1715	1911
1800	2002
1830	2101
1900	2115
2000	2214
2200	0038

4 **a** What time does the last train leave London for Derby?
  b What time does this train arrive in Derby?

Alan notices that the timetable uses the 24-hour clock.

Take away 12 hours from 24-hour clock times to get pm times.

Example	24-hr clock	am/pm
	07 00	7 am (7 o'clock in the morning)
	13 00	1 pm (1 o'clock in the afternoon)
	15 00	3 pm (3 o'clock in the afternoon)
	20 00	8 pm (8 o'clock in the evening)
	23 00	11 pm (11 o'clock in the evening)

Exercise 9:7

1 Write these times as am or pm times.
Remember **am** is morning and **pm** is afternoon and evening.
- **a** 09 00
- **b** 17 00
- **c** 15 45
- **d** 21 35
- **e** 08 05
- **f** 23 00
- **g** 01 15
- **h** 00 30

2 Write these times using the 24-hour clock.
- **a** 4 pm
- **b** 7 pm
- **c** 8.30 pm
- **d** 9.45 pm
- **e** 10.03 pm
- **f** 3.26 pm
- **g** 6 am
- **h** 12.30 pm

3 Here is part of the train timetable.
Write out a new version of this
timetable using am and pm times.

London depart	Derby arrive
1330	1535
1430	1636
1530	1736
1630	1848
1715	1911
1800	2002
1830	2101

4 Here is a timetable using
am and pm times.
Write out a new version
using the 24-hour clock.

London depart	Derby arrive
9.15 am	12.13 pm
12 pm	3.01 pm
2.30 pm	4.49 pm
4.15 pm	6.30 pm
5.40 pm	7.45 pm

Alan wants to work out how long the train takes to get to Derby.

To work out how long the train takes you can use a time line. You work it out in three stages:

1 Count the number of minutes to the next hour.

2 Count the number of full hours.

3 Count the number of minutes to the arrival time.

Example   A train leaves London at 09 25. It arrives in Derby at 11 32. How long does the journey take?

09 25   10 00   11 00   11 32
      35 mins   1 hour   32 mins

The total is **35 mins** + **1 hour** + **32 mins** = 1 hour 67 mins. There are 60 mins in one hour so 67 mins = 1 hour 7 mins.

The journey takes 2 hours and 7 mins.

Exercise 9:8

1   A train leaves London at 08 30. It arrives in Derby at 10 36.
   a Draw a time line to show these times.
   b Work out how long the journey takes.

2   The next train leaves London at 09 30. It arrives in Derby at 11 41.
   a Draw a time line to show these times.
   b Work out how long the journey takes.
   c How much longer is it than the train in question **1**?

3   Look at this part of the timetable.
   a Work out how long each train takes.
   b Which is the quickest train?

London depart	Derby arrive
2.30 pm	4.49 pm
4.15 pm	6.30 pm
5.40 pm	7.45 pm

When Alan gets to Derby he needs to get a bus to Matlock.
Here is the bus timetable from Derby to Matlock.

MONDAYS TO SATURDAYS																	
DERBY, Bus Station (Bay 13)	7.23	–	8.23	–	9.23	–	10.23	–	11.23	–	12.23	–	1.23	–	2.23	–	3.23
Derby, The Strand	7.25	–	8.25	–	9.25	–	10.25	–	11.25	–	12.25	–	1.25	–	2.25	–	3.25
Allestree, Duffield Road/Park Lane	7.34	–	8.34	–	9.34	–	10.34	–	11.34	–	12.34	–	1.34	–	2.34	–	3.34
Duffield, Co-op	7.41	–	8.41	–	9.41	–	10.41	–	11.41	–	12.41	–	1.41	–	2.41	–	3.41
BELPER, Trent Garage	7.51	–	8.51	–	9.51	–	10.51	–	11.51	–	12.51	–	1.51	–	2.51	–	3.51
BELPER, King Street	–	–	–	–	–	–	–	–	–	–	–	–	–	–	–	–	–
John O'Gaunts Way, Whitemoor Centre	–	–	–	–	–	–	–	–	–	–	–	–	–	–	–	–	–
Far Laund	–	–	–	–	–	–	–	–	–	–	–	–	–	–	–	–	–
Heage, Black Boy	–	–	–	–	–	–	–	–	–	–	–	–	–	–	–	–	–
Heage, Heage Tavern	–	–	–	–	–	–	–	–	–	–	–	–	–	–	–	–	–
RIPLEY, Market Place	–	–	–	–	–	–	–	–	–	–	–	–	–	–	–	–	–
Cowers Lane	8.01	–	9.01	–	10.01	–	11.01	–	12.01	–	1.01	–	2.01	–	3.01	–	4.01
Idridgehay	8.09	–	9.09	–	10.09	–	11.09	–	12.09	–	1.09	–	2.09	–	3.09	–	4.09
Wirksworth, St. John's Street	8.16	8.46	9.16	9.46	10.16	10.46	11.16	11.46	12.16	12.46	1.16	1.46	2.16	2.46	3.16	3.46	4.16
Middleton, New Road	8.22	–	–	–	10.22	–	–	–	12.22	–	–	–	2.22	–	–	–	4.22
Cromford, Greyhound	8.33	8.58	9.28	9.58	10.33	10.58	11.28	11.58	12.33	12.58	1.28	1.58	2.33	2.58	3.28	3.58	4.33
Matlock Bath, Fishpond	8.36	9.01	9.31	10.01	10.36	11.01	11.31	12.01	12.36	1.01	1.31	2.01	2.36	3.01	3.31	4.01	4.36
MATLOCK, Bus Station	8.41	9.06	9.36	10.06	10.41	11.06	11.36	12.06	12.41	1.06	1.36	2.06	2.41	3.06	3.36	4.06	4.41

4 **a** What time does the first bus leave Derby?
   b What time does this bus get to Belper, Trent Garage?
   c What time does it get to Wirksworth?
   d What time does it arrive in Matlock?

5 If Alan arrives in Derby at 11.41 am, what is the first bus he can catch to Matlock?

6 The bus that leaves Derby at 1.23 pm arrives in Matlock at 2.41 pm. How long does this journey take?

7 The 2.23 pm bus from Derby arrives at 3.36 pm.
   a How long does this journey take?
   b How much quicker is this than the bus in question **6**?
   c Why do you think this bus takes less time?

8 Some buses only travel between Wirksworth and Matlock. At how many minutes past each hour do these buses leave Wirksworth?

9 Alan gets the 12 30 train from London.
   a What time does he arrive in Derby?
   b Which is the first bus he can get to Matlock?
   c How long does his journey take altogether?

Peak Rail steam train service is open to visitors to Matlock.

Alan decides to go on a steam train.

This calender shows when the service runs (dates that are coloured and/or circled).

CALENDAR

APRIL
M	T	W	T	F	S	S
1	2	3	4	⑤	⑥	⑦
⑧	⑨	10	11	12	13	14
15	16	17	18	19	⑳	21
22	23	24	25	26	27	28
29	30					

MAY
M	T	W	T	F	S	S
		1	2	3	④	5
⑥	7	8	9	10	11	12
13	14	15	16	17	18	19
20	21	22	23	24	25	26
27	28	29	30	31		

JUNE
M	T	W	T	F	S	S
					1	2
3	4	5	6	7	8	9
10	11	12	13	14	15	16
17	18	19	20	21	22	23
24	25	26	27	㉘	㉙	㉚

JULY
M	T	W	T	F	S	S
1	2	3	4	5	6	7
8	9	10	11	12	13	14
15	16	17	18	19	20	21
22	23	24	25	26	27	28
29	30	31				

AUGUST
M	T	W	T	F	S	S
		1	2	3	4	
5	6	7	8	9	10	11
12	13	14	15	16	17	18
19	20	21	22	23	24	25
26	27	28	29	30	㉛	

SEPTEMBER
M	T	W	T	F	S	S
						1
2	3	4	5	6	7	8
9	10	11	12	13	14	15
16	17	18	19	20	㉑	22
23	24	25	26	27	28	29
30						

OCTOBER
M	T	W	T	F	S	S
	1	2	3	4	5	6
7	8	9	10	11	12	13
14	15	16	17	18	19	20
21	22	23	24	25	㉖	27
28	29	30	31			

NOVEMBER
M	T	W	T	F	S	S
				1	2	3
4	5	6	7	8	9	10
11	12	13	14	15	16	17
18	19	20	21	22	23	24
25	26	27	28	29	30	

DECEMBER
M	T	W	T	F	S	S
						1
2	3	4	5	6	⑦	⑧
9	10	⑪	12	13	⑭	⑮
16	17	⑱	19	20	㉑	㉒
㉓	㉔	25	26	27	28	29
30	31					

JANUARY
M	T	W	T	F	S	S
	1	2	3	4	5	
6	7	8	9	10	11	12
13	14	15	16	17	18	19
20	21	22	23	24	25	26
27	28	29	30	31		

FEBRUARY
M	T	W	T	F	S	S
					1	2
3	4	5	6	7	8	9
10	11	12	13	14	15	16
17	18	19	20	21	22	23
24	25	26	27	28		

MARCH
M	T	W	T	F	S	S
					1	2
3	4	5	6	7	8	9
10	11	12	13	14	15	16
17	18	19	20	21	22	23
24	25	26	27	㉘	㉙	㉚
㉛	Ⓐⓟⓡ					

Key:

⬜ Low Season

⬛ High Season

▨ Midwest/Winter Season

◺ Diesel Day

◯ Special Event Day

The timetable shows when the trains depart (d) from and arrive (a) at Darley Dale and Matlock.

TIMETABLE

Darley Dale (d)		10.00	10.30	11.30	12.30	13.30	14.30	15.30
Matlock (a)		10.10	10.40	11.40	12.40	13.40	14.40	15.40
Matlock (d)	09.45	10.15	10.45	11.55	12.55	13.55	14.55	15.55
Darley Dale (a)	09.55	10.25	10.55	12.05	13.05	14.05	15.05	16.05

Exercise 9:9

1   How many days in August do the trains run?

2   Which month has the most Special Event Days? (dates are circled)

3   There are two days each week in July when the train does not run. Which days are they?

4   Which two months have the fewest running days in them?

5   Which months are partly High season?

6   Write down the dates of all the 'Diesel Days' in the year.

7   **a** What time does the first train leave Matlock in the morning?
    b What time does this train arrive in Darley Dale?
    c How long does the journey take?

8   Alan took the 10 45 train to Darley Dale.
    He caught the 14 30 back to Matlock.
    How long did he spend at Darley Dale?

9   What time does the last train leave Darley Dale?

Here is another copy of the timetable, including times for Sunday only (SuO).

									SuO
Darley Dale (d)		10.00	10.30	11.30	12.30	13.30	14.30	15.30	16.30
Matlock (a)		10.10	10.40	11.40	12.40	13.40	14.40	15.40	16.40
Matlock (d)	09.45	10.15	10.45	11.55	12.55	13.55	14.55	15.55	16.55
Darley Dale (a)	09.55	10.25	10.55	12.05	13.05	14.05	15.05	16.05	17.05

The outside pink box means that all of the trains run in the High Season.
The trains in the inside green box run in the Low Season.
The trains in the blue box are the only ones that run in the winter and midweek.

Exercise 9:10

1 Does the 10 30 train from Darley Dale run in the winter?

2 In which season does the 16 30 from Darley Dale run?

3 In which season is 8th September?

4 Do any trains run on November 5th?

5 Jim arrives in Darley Dale at 11 07 on a Sunday in February.
    a In which season are the Sundays in February?
    b What is the first train Jim can catch to Matlock?
    c How long will he have to wait for the train?

6 Anne catches the 11.55 am train from Matlock.
    She wants to get back to Matlock by 3.25 pm.
    Which train should she catch from Darley Dale?

7 David visits Matlock from 29th July until 5th August.
    a How many days do the trains run in this time?
    b How many of these days are High Season?
    c David catches the first train from Matlock to Darley Dale on
        3rd August. What time does this train leave Matlock?
    d David catches the last train back in the afternoon.
        What time does this train leave Darley Dale?
    e How long does David spend in Darley Dale?

1   **a** Write down all the months in a year with 31 days.

     b How many days are there in a leap year?

     c Which of these years are leap years?

         2001, 2006, 2010, 2016, 2019, 2024

2 How long does this film
last in hours and minutes?

3   **a** Carl is keen on a computer game called Minesweeper.
His best time is 412 seconds.
What is Carl's best time in minutes and seconds?

     b Carl's sister Sian says her best time is 6 minutes 35 seconds.
What is Sian's time in seconds?

     c Who is faster and by how many seconds?

4 Write these times in words.

   Example   3.45 pm is a quarter to four in the afternoon.

   a 5.10 pm   **c** 10.47 am   **e** 6.25 am   **g** 12.42 pm
   b 7.35 am   **d** 2.54 pm   **f** 9.38 pm   **h** 8.09 pm

5 Sally is cooking chicken tikka.
The time is ten to six.

     a When will the chicken tikka be done if
Sally uses her microwave?

     b Her microwave is broken so Sally uses
her conventional oven.
The oven takes 15 minutes to heat up.
When will the chicken tikka be done?

Chicken tikka

Microwave on full power
for five minutes.

Conventional oven: Put in
a pre-heated oven at 200 °C
for 25 minutes.

Chicken tikka

6 Gary's recipe for roast turkey says to cook for
35 minutes per kilogram and 30 minutes extra.
Gary's turkey weighs 8 kilograms.

     a How long does the turkey take to cook in minutes?

     b How long does the turkey take to cook in hours and minutes?

     c The turkey goes in the oven at 2.45 pm.
When will it be done?

7 a Look at this train timetable.
What time does the first train leave Bristol?

Bristol	Leeds
0915	1213
1200	1501
1430	1649
1615	1830
1740	1945
1800	2026
1830	2101
1915	2134
2045	2300

b What time does this train arrive in Leeds?
c What time does the 17 40 from Bristol arrive in Leeds?
d Nick arrives at Bristol station at 1.15 pm.
He wants to go to Leeds.
What time does the next train leave?
e Mariza catches the 19 15 train to Leeds.
How long does her journey take?

8 Terry's school bus is due at 08 12.
a The bus is 4 minutes late.
What time does the bus arrive?
b Terry got to the bus stop at 08 03.
How long did he have to wait for the bus?
c Terry's bus ride takes 18 minutes.
What time does Terry arrive at school?

9 Here is a bus timetable.

Cubley	08 25	08 55	09 15	09 25	09 50	10 05
Ashbourne	08 45	09 13	09 24	10 00	10 13	10 29

a Rajdeep catches the first bus from Cubley.
What time does she arrive in Ashbourne?
b One bus leaves Cubley at 08 55.
How long does it take to get to Ashbourne?
c Robbie is 6 minutes late for the 09 50 to Ashbourne.
How long does he have to wait for the next bus?

10   While Pete is in Matlock, he and Sally have a day out in Manchester.
They catch the TransPeak bus to Manchester.
Here is the timetable.

DERBY TO MANCHESTER	MORNINGS			
DERBY, Bus Station (bay 13) dep.	–	7.55	9.55	11.55
Belper, Trent Garage	–	8.15	10.15	12.15
Matlock Bath, Fishpond Hotel	–	8.30	10.30	12.30
MATLOCK, Bus Station (bay 1)	–	8.37	10.37	12.37
Hackney Lane	–	8.42	10.42	12.42
Darley Dale, Post Office	–	8.46	10.46	12.46
Rowsley, Peacock Hotel	–	8.50	10.50	12.50
BAKEWELL, Rutland Arms	–	8.59	10.59	12.59
Ashford-in-the-Water, Post Office	–	9.04	11.04	1.04
Taddington Village	–	9.13	11.13	1.13
BUXTON, Market Place	6.51	9.28	11.28	1.28
Dove Holes, Kays Truck Centre	7.02	9.37	11.37	1.37
Chapel-en-le-Frith, New Inn	7.07	9.42	11.42	1.42
Whaley Bridge, Railway Station	7.14	9.49	11.49	1.49
New Mills, The Swan	7.21	9.56	11.56	1.56
Disley, Rams Head Hotel	7.23	9.58	11.58	1.58
Hazel Grove, Rising Sun	7.33	10.08	12.08	2.08
Stockport, Mersey Sq. (opp. Debenhams)	7.48	10.23	12.23	2.23
MANCHESTER, Chorlton St. Coach Stn.	8.20	10.45	12.45	2.45

MANCHESTER TO DERBY	AFTERNOONS		
MANCHESTER, Chorlton St. Coach Stn. (bay E2)	11.15	1.15	3.15
Stockport, Bus Station (platform B)	11.37	1.37	3.37
Hazel Grove, Rising Sun	11.52	1.52	3.52
Disley, Rams Head Hotel	12.02	2.02	4.02
New Mills, The Swan	12.04	2.04	4.04
Whaley Bridge, Railway Station	12.11	2.11	4.11
Chapel-en-le-Frith, New Inn	12.18	2.18	4.18
Dove Holes, Kays Truck Centre	12.24	2.24	4.24
BUXTON, Market Place	12.34	2.34	4.34
Taddington Village	12.49	2.49	4.49
Ashford-in-the-Water, Post Office	12.57	2.57	4.57
BAKEWELL, Rutland Arms	1.03	3.03	5.03
Rowsley, Peacock Hotel	1.12	3.12	5.12
Darley Dale, Post Office	1.16	3.16	5.16
Hackney Lane	1.20	3.20	5.20
MATLOCK, Bus Station (bay 6)	1.25	3.25	5.25
Matlock Bath, Fishpond Hotel	1.30	3.30	5.30
Belper, Chapel St outside Trent Garage	1.45	3.45	5.45
DERBY, Bus Station (bay 18)	2.05	4.05	6.05

a   (1) What time does the first bus leave Matlock in the morning?
    (2) What time does this bus get to Manchester coach station?

b   Pete and Sally catch the 8.37 am bus from Matlock.
    (1) What time does this bus get to Buxton?
    (2) How long does it take to get from Matlock to Buxton?
    (3) What time does it arrive in Manchester?
    (4) How long is the journey altogether?

c   Pete and Sally arrive in Manchester at 10.45 am.
    Their bus home leaves at 5.20 pm.
    How long can they spend in Manchester?

d   They decide to get the 1.15 pm bus from Manchester.
    (1) What time does this bus get to Buxton?
    (2) How long does this journey to Buxton take?
       Pete and Sally get off in Buxton to look around.
    (3) What time does the next bus leave Buxton for Matlock?
    (4) How long can Pete stay in Buxton?
    (5) What time will they get back to Matlock?

1   **a**   The 6th June is a Tuesday.
       Write down the number pattern of all the Tuesdays in June.
     b   The 4th May is a Thursday.
       Write down the number pattern of all the Thursdays in May.
     c   The 26th July is a Wednesday.
       Write down the number pattern of all the Wednesdays in July.

2   Katie is putting a birthday cake into the oven to bake.
     The time is 5.35 pm.
     The cake needs three and three quarter hours to cook.
     When does Katie need to take the cake out of the oven?

3   These are the afternoon trains
from London to Derby.
Which train takes the most time
for its journey?

London	Derby
1230	1440
1330	1535
1430	1636
1530	1736
1630	1848
1715	1911

4   Here is a bus timetable.

Billingham	08 15	08 45	09 15
Norton	08 32	09 04	09 27
Stockton	08 41	09 20	09 39
Thornaby	08 55	09 34	09 51

     a   How long does the 08 45 from Billingham take to get to Thornaby?
     b   Another bus leaves Billingham for Thornaby at 09 15.
       Does this bus take less time than the 08 45?
     c   Derek gets to the bus stop in Billingham 13 minutes late for the
       08 15 bus.
       How long does he wait for the next bus?
     d   Derek catches the 08 45 from Billingham.
       He works in Stockton.
       Derek's work is 10 minutes walk from the bus stop.
       What time does Derek arrive at work?

- 1 ordinary year = 365 days
  1 leap year = 366 days
  Leap years are …, 2000, 2004, 2008, 2012, 2016, …
  Times after midnight and in the morning are am.
  Times in the afternoon and evening are pm.

- *Example*   Sally's mother is cooking this chicken.
  It takes 1 hour 45 minutes to cook.
  She puts the chicken in the oven
  at 9.50 am. When will it be cooked?

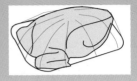

Sort the 45 minutes   At 9.50 am it is **10 minutes**
out first.   to the next hour.
   45 − 10 = 35

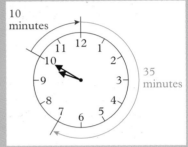

*If the chicken took **45** minutes it would be ready*
at 10.35 am.

10.35 + 1 hour = 11.35 am

The chicken will be cooked at 11.35 am.

- Take away 12 hours from 24-hour clock times to get pm times.

Examples

24-hr clock	am/pm
13 00	1 pm (1 o'clock in the afternoon)
15 00	3 pm (3 o'clock in the afternoon)
23 00	11 pm (11 o'clock in the evening)

- To work out how long a train takes to arrive at its destination you can use a
  time line. You work it out in three stages:
  1 Count the number of minutes to the next hour.
  2 Count the number of full hours.
  3 Count the number of minutes to the arrival time.

Example   A train leaves London at 09 25. It arrives in Derby at 11 32.
   How long does the journey take?

The total is **35 mins** + 1 hour + **32 mins** = 1 hour 67 mins.
There are 60 mins in one hour so 67 mins = 1 hour 7 mins.

The journey takes 2 hours and 7 mins.

1   Work out the number of
    a   seconds in four minutes   **c**   days in eight weeks
    b   minutes in five hours   **d**   months in four years.

2   The 1st March is a Wednesday.
    Write down the number pattern of all the Wednesdays in March.

3   Write these times as am or pm times.
    a   Ten past three in the afternoon.
    b   Twenty-five to four in the afternoon.
    c   A quarter past midnight.
    d   Five to seven in the morning.

4   Work out the times when each of these will be cooked.
    a   A pizza is put in the oven at 5.42 pm.
        It needs 25 minutes.
    b   An Irish stew is put in the oven at 3.50 pm.
        It needs 2 hours 15 minutes.

5   Write these times as am or pm times.
    Remember **am** is morning and **pm** is afternoon and evening.
    a   10 23   **b**   14 25   **c**   16 48   **d**   22 47

6   A train leaves Nottingham at 10 15. It arrives in Durham at 12 27.
    a   Draw a time line to show these times.
    b   Work out how long the journey takes.

7   Here is a bus timetable:

Krustleton	08 15	08 30	08 45	09 00
Barbland	08 35	08 50	09 05	09 20
Ren Hill	08 50	09 05	09 20	09 35

    a   What time does the first bus leave Krustleton?
    b   What time does this bus arrive in Ren Hill?
    c   How long does it take for its journey?
    d   Paul gets to the bus stop in Krustleton at 08 32.
        How long does he have to wait for his bus?

10 Negative numbers

CORE

1 **Temperature**

2 **Other number scales**

QUESTIONS

EXTENSION

SUMMARY

TEST YOURSELF

Under water, the deeper you go, the greater the weight and the pressure of water above you.

Nitrogen from an air supply is dissolved inside a diver's body under the high pressure.

If the diver comes back up too fast the nitrogen is released as tiny bubbles in the bloodstream. These bubbles can block the oxygen supply to tissues and organs causing dizziness and cramps – a condition known as 'the bends'.

1 Temperature

The Voyager spacecraft is visiting the largest of the planets, Jupiter.
The average temperature on Jupiter is $-120\,°C$.

We usually use the Celsius scale to measure temperature.
On the Celsius scale, the freezing point of water is $0\,°C$
and the boiling point is $100\,°C$.
Temperatures lower than freezing on the Celsius scale
are written with a **minus sign in front**.
They are **negative numbers**.

Examples   Negative numbers $-120, -50, -1$
Positive numbers $3, +45, 100$
(We do not always use $+$, so $+3$ is the same as 3.)
Nought is not positive or negative.

Exercise 10:1

1   From this list: $-60, 10, 17, +85, 0, -273, -5$
  a Write down the negative numbers.
  b Write down the positive numbers.
  c Which number is not positive or negative?

2 Write down the temperatures in degrees Celsius (°C) shown by these thermometer scales.

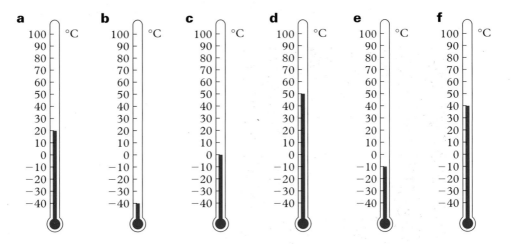

Less than	< means 'less than' or, for temperature, 'colder than'.
More than	> means 'more than' or, for temperature, 'warmer than'.
Examples	−5°C < −2°C means −5°C is less than or colder than −2°C
	2°C > −4°C means 2°C is more than or warmer than −4°C

3 Is each of these true or false?

a −7°C < −5°C   **c** 6°C > −1°C   **e** −100°C > −30°C

b −9°C > −4°C   **d** −50°C < −10°C   **f** 15°C < 20°C

4 Put these temperatures in order, coldest first.

a −6°C, 5°C, −8°C, 0°C, 1°C

b −1°C, −2°C, 3°C, 4°C, 10°C

c −5°C, 3°C, −7°C, 6°C, −10°C

5 Put these temperatures in order, warmest first.

a 4°C, −1°C, 9°C, −5°C, 11°C

b 3°C, −10°C, 8°C, −4°C, 2°C

c 9°C, −4°C, −3°C, 6°C, −12°C

6 Earth and the other planets are warmed by the Sun.
Here is a list of planets with their average temperatures.

a Write the list in order, warmest at the top.

Earth	15°C
Jupiter	−120°C
Mars	−40°C
Mercury	13°C
Neptune	−220°C
Pluto	−250°C
Saturn	−180°C
Uranus	−210°C
Venus	460°C

b Which planet is the hottest?
c Which planet is the coldest?

7 What is the difference between these day and night temperatures?
The first one has been done for you.

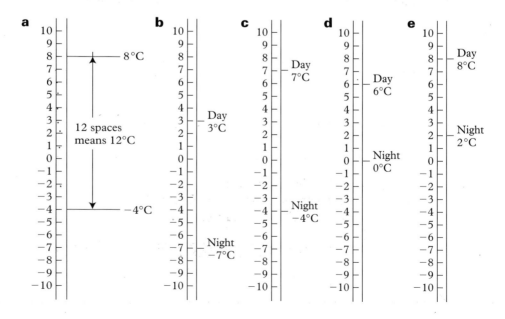

8 This bar-chart shows the mean temperature for each month in Moscow.

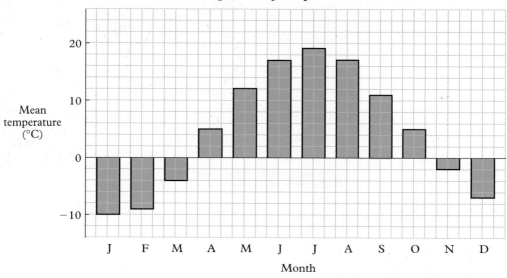

Average monthly temperatures in Moscow

a What is the mean temperature during the hottest month?
b What is the mean temperature during the coldest month?
c What is the difference between these mean temperatures?

You can use a calculator to help you with temperatures.

Example

One day the temperature is 3°C.
That night the temperature falls by 7°C.
Find the night temperature.

Keys to press: **3** **−** **7** **=** -4

Answer −4°C

Exercise 10:2

1 Find the night time temperature.

Daytime temperature	Fall in temperature	Night time temperature
a 4°C	7°C	4 − 7 = ...
b 14°C	2°C	
c 8°C	12°C	
d 10°C	8°C	
e 6°C	14°C	

Example   One day the temperature is 9°C. That night the temperature is −3°C. What is the difference between the day and night temperatures?

Keys to press: **9** **−** **3** **+/−** **=** 12

Answer 12°C

---

2 Copy the table and fill it in.
Find the difference between each pair of day and night temperatures.

Day temperature	Night temperature	Keys to press	Difference
a 6°C	−4°C	**6** **−** **4** **+/−** **=**	...°C
b 1°C	−7°C		
c 3°C	−10°C		
d 5°C	−3°C		
e 9°C	2°C		
● **f** −1°C	−6°C		

3 What is the difference between the mean temperature on Earth, 15°C, and the mean temperature on Mars, −40°C?

4 The mean temperature on the side of Mercury that faces the Sun is 430°C. The mean temperature on the dark side of Mercury is −170°C. What is the difference in temperature between the two sides?

5 The hottest planet is Venus.
The mean temperature on Venus is 460°C.
The coldest planet is Pluto.
The mean temperature on Pluto is −250°C.
What is the difference in temperature between the two planets?

2   Other number scales

In a TV quiz game both sides start with no points.
Ken interrupts the first question. He gets the wrong answer.
Ken's team loses 5 points.
Ken's team now have a score of −5.

Exercise 10:3

1 The scoreboard shows a team's
score in a quiz game.
The team now gets 10 points.
What is their new score?

$$-5$$

2 Here is the control panel of a lift
in an office block.

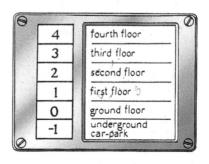

4	fourth floor
3	third floor
2	second floor
1	first floor
0	ground floor
-1	underground car-park

  a What number is used for the
  ground floor?
  b The boiler room is on the
  first floor.
  What number is the boiler
  room?
  c Where do you go if you press
  −1?
  d Katie goes from the basement to
  the third floor.
  How many floors does she go up?
  e The caretaker goes from the
  second floor to the boiler room.
  How many floors does he go
  down?

3 Each of these number lines has two missing numbers.
Write down the missing numbers for each line.

a

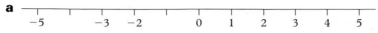

b

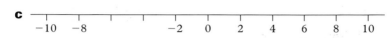

c

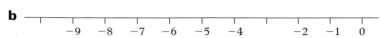

d

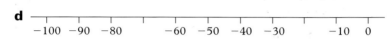

4 a Copy this number line.

b Label the line from −6 to 6.

Example   Find the rule for this number pattern.
Write down the next two terms.
4, 2, 0, −2, −4, ..., ...

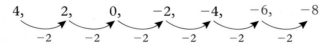

The rule is −2.
The next two terms are −6 and −8.

---

5 Write down the rule for each number pattern.
Write down the next two terms.
a 5, 3, 1, −1, −3, ..., ...   **c** 50, 40, 30, 20, 10, ..., ...
b 6, 3, 0, −3, −6, ..., ...   **d** 0, −5, −10, −15, −20, ..., ...

6 The diagram shows the cross-section of a mine. The scale is in metres.

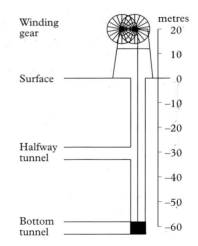

  a How far is it from the halfway tunnel to the surface?

  b How far does the lift cage travel from the surface to the bottom tunnel?

  c How far is it from the winding gear to the halfway tunnel?

  d How far is it from the winding gear to the bottom tunnel?

  e How far is it from the bottom tunnel to the halfway tunnel?

7 The helicopter is lowering a man down to the lighthouse. The scale is in metres.

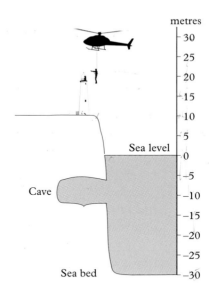

  a How high is the man above the top of the cliff?

  b How many metres is the helicopter above the bottom of the sea?

  c How many metres is the top of the lighthouse above the sea bed?

  d How many metres is the cave above the sea bed?

  e How many metres above the cave is the top of the cliff?

1 **a** Which temperature is warmer $-10°C$ or $5°C$?
    b Which temperature is warmer $-20°C$ or $-30°C$?
    c Which temperature is warmer $-80°C$ or $-15°C$?
    d Which temperature is warmer $-100°C$ or $100°C$?

2 **a** Which temperature is colder $-30°C$ or $25°C$?
    b Which temperature is colder $60°C$ or $10°C$?
    c Which temperature is colder $5°C$ or $-15°C$?
    d Which temperature is colder $-17°C$ or $-35°C$?

3 Place a $<$ or a $>$ between each pair of temperatures.
    a $3°C$ $8°C$ **c** $-15°C$ $-50°C$ **e** $75°C$ $120°C$
    b $-7°C$ $-10°C$ **d** $20°C$ $-13°C$ **f** $-90°C$ $-40°C$

4 Use your calculator to work these out.
    a $2-4$ **c** $4-5$ **e** $0-3$ **g** $10-5$
    b $7-8$ **d** $0-5$ **f** $1-2$ **h** $5-10$

5 One day the temperature is $8°C$.
    That night the temperature falls by $10°C$.
    What is the night time temperature?

6 Jon puts a chicken at $20°C$ into his freezer which is at $-18°C$.
    How many degrees will the chicken's temperature drop?

7 Find the difference between these day temperatures and the night
    temperatures.

Day	Night	Difference
$7°C$	$-2°C$	
$4°C$	$-8°C$	
$0°C$	$-12°C$	
$10°C$	$-1°C$	

8 Write these temperatures in order, coldest first.
    $1°C, -5°C, -17°C, 11°C, 14°C$

9 Brian has drawn a picture of himself fishing in a river. The scale is in metres.

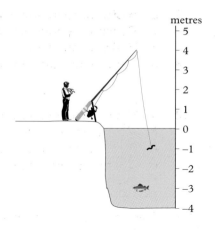

a How high is the top of Brian's fishing rod from the water?

b How far is the fish below the surface of the water?

c How far is it from the top of Brian's rod to the bottom of the river?

d How far is it from the worm to the bottom of the river?

10 a Copy the number line shown and the outline of the cliff in the picture.
The distances are in metres.

b Add these to your diagram:

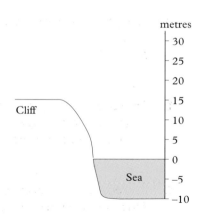

Description	Height
house	15 m
seagull	25 m
birds' nest	10 m
crab	−10 m
small boat	0 m
fish	−5 m

c How far below the seagull is the fish?

d How far above the crab is the birds' nest?

e How far above the crab is the fish?

1   Look at the table.
It gives the mean temperature each month at Beijing.

Month	J	F	M	A	M	J	J	A	S	O	N	D
Temp °C	−5	−2	4	13	20	24	26	24	19	13	3	−3

a   What is the mean temperature during the hottest month?
b   What is the mean temperature during the coldest month?
c   What is the difference between these mean temperatures?
d   Which months have a mean temperature below zero?
e   Copy these axes on to squared paper.

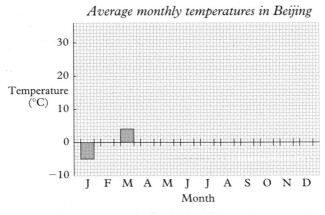

Average monthly temperatures in Beijing

f   Draw a bar-chart to show the temperatures.

2   Look at this time line.
a   When was the Forth
Rail Bridge built?
b   When were the
Pyramids of Egypt
built?
c   How many years after
the Great Wall of China
was Hadrian's Wall
built?
d   How many years after
Hadrian's Wall was the
Forth Rail Bridge built?
e   How many years before
the Great Wall of China
were the Pyramids of
Egypt built?

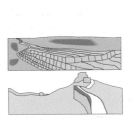

2000 AD
1890 AD Forth Rail Bridge

1000 AD

125 AD Hadrian's Wall
0
240 BC Great Wall of China

1000 BC

2000 BC

2600 BC Pyramids of Egypt

3000 BC

- We usually use the Celsius scale to measure temperature.
  On the Celsius scale, the freezing point of pure water is $0°C$ and its boiling point is $100°C$.
  Temperatures lower than freezing on the Celsius scale are written with a **minus sign in front**.
  They are **negative numbers**.

  Examples   Negative numbers $-120, -50, -1$
  Positive numbers $3, +45, 100$
  (We do not always use $+$, so $+3$ is the same as 3.)
  Nought is not positive or negative.

- **Less than** $<$ means '**less than**' or '**colder than**' for temperature
  More than $>$ means '**more than**' or '**warmer than**' for temperature

  Examples $-5°C < -2°C$ means $-5°C$ is less than or colder than $-2°C$
  $2°C > -4°C$ means $2°C$ is more than or warmer than $-4°C$

- You can use a calculator to help you with temperatures.

  Example   One day the temperature is $3°C$.
  That night the temperature falls by $7°C$.
  Find the night temperature.

  Keys to press: **3**   **−**   **7**   **=** -4
  Answer $-4°C$

- *Example*   One day the temperature is $9°C$.
  That night the temperature is $-3°C$.
  What is the difference between the day and night temperatures?

  Keys to press: **9**   **−**   **3**   **+/−**   **=** 12
  Answer $12°C$

- *Example*   Find the rule for this number pattern.
  Write down the next two terms.
  $4, 2, 0, -2, -4, ..., ...$

  The rule is -2.
  The next two terms are $-6, -8$.

1 Place a $<$ or a $>$ between each pair of temperatures.
   a $2°C$ $-7°C$ **b** $-12°C$ $-3°C$ **c** $-1°C$ $8°C$

2 Write these temperatures in order, coldest first.
   $-5°C, 0°C, 15°C, 4°C, -9°C$

3 One day the temperature was $8°C$.
   That night the temperature fell by $12°C$.
   What was the night time temperature?

4 Copy the table and fill it in.
   Find the difference between each pair of day and night temperatures.

Day temperature	Night temperature	Keys to press	Difference
$5°C$ $4°C$	$-6°C$ $-3°C$		

5 Work these out on a calculator.
   a $11 - 13$ **b** $0 - 4$ **c** $15 - 13$ **d** $1 - 5$

6 The mean temperature of Jupiter is $120°C$.
   The mean temperature of Earth is $15°C$.
   What is the difference between these temperatures?

7 Write down the rule for each number pattern.
   Write down the next two terms.
   a $1, 0, -1, -2, -3, ..., ...$ **b** $20, 16, 12, 8, 4, ..., ...$

8 Duncan has drawn this picture of
   the seaside.
   The heights on the scale are in
   metres.
   a How far below sea level is the
      jellyfish?
   b How many metres above the
      bottom of the sea is the jellyfish?
   c How many metres above the
      bottom of the sea is the seagull?
   d How many metres above the
      bottom of the sea is the top of
      the sand dune?

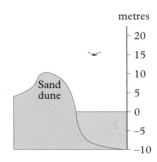

Michelle Smith won three gold medals for Ireland in swimming events at the 1996 Olympics in Atlanta:

200-metres individual medley

400-metres individual medley

400-metres freestyle

1   Olympic standard averages

This is the women's 100 m final from the Olympic Games in Atlanta.
What makes a good sprinter?
Should they be tall and light?
What is the average sprinter like?

Mean

To find the **mean** of a set of data:
a   find the total of all the data values
b   divide the total by the number of data values.

Example

Here are the weights of 6 sprinters in kilograms.
Find their mean weight.

The total is:

$$88 + 90 + 79 + 94 + 86 + 91 = 528 \text{ kg}$$

The mean is $528 \div 6 = 88$ kg.

Exercise 11:1

1   Here are the weights of 5 sprinters in kilograms.
   a   Find the total weight of the sprinters.
   b   Find their mean weight.

   85   91   74   68   82

2   These are the heights in centimetres of the same 5 sprinters.
   a   Find the total of the heights.
   b   Find their mean height.

   170   190   185   178   188

3 These are the times the sprinters took to run a 100 m race.
The times are in seconds.

     10.5   10.7   10.1   9.9   10.3

  a The scale shows the time taken in seconds. Copy the scale.
     The gap between each mark is 1 cm.

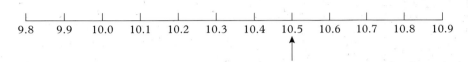

  b Put an arrow below each time on your scale.
     The first one has been done for you.
  c Find the total of the 5 times.
  d Find the mean time taken.
  e Mark the mean time on your scale. Use a new colour.
  f Is the mean time one of the times of the runners?

4 These are the weights of 5 of the
weightlifters at the Olympics.
They are in kilograms.

     104   108   112   107   104

  a Find the total weight of these men.
  b Find the mean weight.
  c Do you think these men would
     make good sprinters?
     Explain your answer.

5 These are the heights of the weightlifters.
The heights are in centimetres.

     168   188   187   175   182

  a Find the mean height of the weightlifters.
  b Compare this height with the answer to question **2**.
     Is there a big difference?

Exercise 11.2

1 This table shows the medals won at the Olympic Games by the USA.

Year	Gold	Silver	Bronze	Total
1996	42	32	25	99
1992	37	33	37	107
1988	35	31	27	93

   a How many gold medals did the USA win in 1992?
   b How many bronze medals did they win in 1988?
   c How many gold medals have they won altogether in the last three Olympics?
   d Find the mean number of gold medals won in the last three Olympics.
   e Find the mean number of silver medals won in the last three Olympics.

2 This is the medal table for Germany.

Year	Gold	Silver	Bronze	Total
1996	19	17	27	63
1992	33	21	28	82
1988	37	35	30	102

   a How many gold medals did Germany win in 1992?
   b How many silver medals did they win in 1996?
   c How many gold medals have they won altogether in the last three Olympics?
   d Find the mean number of gold medals won in the last three Olympics.
   e Find the mean number of silver medals won in the last three Olympics.
   f Find the mean number of bronze medals won in the last three Olympics.

Another type of average is the **mode**.
This is often used when the average has to be whole number.
It is easy to find.

Mode	The **mode** is the most common or most popular data value. It is sometimes called the **modal value**.

Example

These are the winning distances in the men's javelin for the past few Olympics. They are given to the nearest metre. Find the modal distance.

90   90   95   91   89   84   90

There have been more **90** m winning throws than anything else. The mode is **90** m.

Exercise 11:3

1   These are the silver medal throws in the men's javelin.
They are rounded to the nearest metre.

89   90   88   90   86   84   87

What is the mode of these distances?

2   These are the bronze medal throws in the men's javelin.
They are rounded to the nearest metre.

87   84   87   87   84   83   83

What is the mode of these distances?

● **3**   The distances in the men's javelin in 1996 were:

Gold: 88 m
Silver: 87 m
Bronze: 87 m

If you include these distances with the previous ones, would any of the modes change?

4 Here are some of the results for the women's javelin.
They have been rounded to the nearest metre.

Year	Gold	Silver	Bronze
1968	60	60	58
1972	64	63	60
1976	66	65	64
1980	68	68	67
1984	70	69	67
1988	75	70	67
1992	68	68	67

  a How far was the winning throw in 1968?
  b How much further than the 1968 distance was the winning throw in 1988?
  c What is the modal distance to win a gold medal?
  d What is the modal distance to win a silver medal?
  e What is the modal distance to win a bronze medal?
  f The women's results for 1996 were:

    Gold: 68 m
    Silver: 66 m
    Bronze: 65 m

    Do these results change any of the modes?

5 These pie-charts show the medals won by the UK and Poland in the 1996 Olympics. You can't see how **many** medals they won.

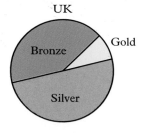

  a What is the modal type of medal for the UK?
  b What is the modal type of medal for Poland?
  c Explain how you found the answer.

Sometimes the mode is not a useful average to find. The most common data value could be the smallest or the biggest number. It is often more useful to have a middle value.

There is a third type of average. It is called the **median**.

Median

To find the median you put all the data values in order of size. The **median** is then the middle value.

Example

The number of medals won by Japan at the Olympics from 1972–1996 are:

29   25   0   32   14   22   14

Find the median number of medals Japan has won.

Write the numbers in order. Start with the smallest.

0   14   14   (22)   25   29   32

Find the middle number. It is **22**.
The median number of medals is **22**.

Exercise 11:4

1   Here are the number of Olympic medals won by Italy from 1972–1996.

18   13   15   32   14   19   34

   a   Write these numbers in order.
      Start with the smallest.
   b   Find the median number of medals
      won by Italy.

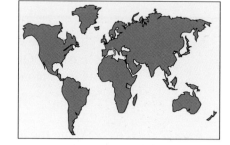

2   Here are the number of Olympic medals won by Kenya from 1972–1996.

9   0   0   3   9   8   8

   a   Write these numbers in order.
      Start with the smallest.
   b   Find the median number of medals
      won by Kenya.

3 Here are the number of Olympic medals won by Sweden from 1972–1996.

16   5   12   19   11   12   7

a Find the median number of medals won by Sweden.
b What is the modal number of medals?

4 Here are the number of Olympic medals won by Australia from 1972–1996.

17   5   9   24   14   27   40

a Find the median number of medals won by Australia.
b Find the total number of medals they have won.
c Divide your answer to **b** by 7 to find the mean.
d Explain why you can't find the mode.

If you have an even number of data values then there will not be one exactly in the middle.
When this happens find the mean of the middle two numbers.

Example

These are the winning distances in the men's javelin for the past few Olympics. They are given to the nearest metre. Find the median distance.

95   91   89   84   90   88

Write the numbers in order.

84   88 | 89   90 | 91   95

Find the middle two numbers: **89**   **90**

Find the mean: $\dfrac{89 + 90}{2} = 89.5$

The median distance is 89.5 m.

---

5 These are the winning distances in the men's discus for the past few Olympics. They are given to the nearest metre.

61   65   64   68   67   67   69   65

a Write these numbers in order.
   Start with the smallest.
b Write down the **two** middle numbers.
c Add these numbers together and divide by 2.
   This is the median distance.

6 These are the silver medal distances in the men's discus for the past few Olympics. They are given to the nearest metre.

61   63   64   66   66   66   67   65

a Write these numbers in order.
   Start with the smallest.
b Write down the **two** middle numbers.
c Find the median distance.

7 These are the winning times in the women's 100 m for the past few Olympics. They are given to the nearest tenth of a second.

11.4   11.0   11.1   11.1   11.1   11.0   10.5   10.8

a Write these numbers in order. Start with the smallest.
b Write down the **two** middle numbers.
c Find the median time.

8 These are the bronze medal distances in the men's discus for the past few Olympics. They are given to the nearest metre.

59   63   63   66   66   65   67   64

Find the median distance.

9 These are the silver medal times in the women's 100 m for the past few Olympics. They are given to the nearest tenth of a second.

11.6   11.1   11.2   11.1   11.1   11.1   10.8   10.8

Write down the median time.

2 Solving problems with statistics

Cars travel a total of 270 000 million kilometres in the UK every year.
The average for the UK, France, Germany and Spain is 255 000 million
kilometres.

This average does not tell the whole story.
There is a big difference between the countries.

Germany's total is the biggest at 376 000.
Spain's total is the smallest, at only 70 000.

The difference of 306 000 million kilometres is called the **range** of the data.

Range   The **range** of a set of data is the biggest value take away the
smallest value.

Exercise 11:5

1 These are the numbers of people hurt in road accidents in different
countries over a four year period.

Country	Number of people injured
Germany	448 000
France	244 000
UK	313 400
Spain	166 000
Iceland	939
Norway	10 900

a Which country had the biggest number of accidents?
Write down the number of accidents.
b Which country had the smallest number of accidents?
Write down the number of accidents.
c Work out the biggest number take away the smallest.
This is the range.

2 These are the numbers of passengers using each country's biggest
airport over the same four year period.

Country	Number of passengers
Germany	24 300 000
France	22 200 000
UK	37 500 000
Spain	13 200 000
Iceland	532 000
Norway	6 000 000

a Which country had the most passengers?
Write down the number of passengers for this country.
b Which country had the least passengers?
Write down the number of passengers for this country.
c Find the range of this data.

3 These are the average ages that people live to in different countries.
This is called Life Expectancy.

Country	Life Expectancy	
	Men	Women
UK	72	78
Angola	43	46
Brazil	62	68
Kenya	57	61
Iceland	75	80
Turkey	63	66

a Work out the range for men.
b Work out the range for women.

The **range** can help you to compare two sets of data.
Two sets of data might have the same mean but their ranges can be different.

Example

Here are the runs scored by two cricketers in their last six innings.

Ian	44	73	39	60	68	40
Mike	120	7	84	26	9	90

Ian's mean is 54 runs. His range is $73 - 39 = 34$ runs.
Mike's mean is 56 runs. His range is $120 - 7 = 113$ runs.

Their means are very similar but Mike's range is much bigger.
The bigger range shows that Mike may score a lot of runs but he may score very few.
Ian's smaller range means that he is more steady and reliable.

If you needed 50 runs to win a match Ian would be the best choice.
If you needed 100 runs to win a match Mike would have the best chance.

Exercise 11:6

1 Lindsey uses two different routes to drive to work.
She times her journeys for 5 days on each route to the nearest minute.
Here are her results.

Route 1	43	50	47	53	42
Route 2	39	62	47	65	37

  a Find Lindsey's mean time for route 1.
  b Find Lindsey's mean time for route 2.
  c Find the range of Lindsey's times for route 1.
  d Find the range of Lindsey's times for route 2.
  e Lindsey has to get to work for an important meeting in 50 minutes time.
  Which would be her best route? Explain your answer.

2 Derek also has two routes to travel to work. Here are his times.

Route 1	15	24	19	26	11
Route 2	18	19	18	19	21

Derek has to be at work in 20 minutes time.
Which would be the best route to take? Explain your answer.

Exercise 11:7

Look at this data. It is about 10 pupils in Year 11.
The data was taken when they were in Year 7, Year 9 and Year 11.
The heights are to the nearest centimetre.

Pupil	Sex	Y7 height	Y7 shoe size	Y9 height	Y9 shoe size	Y11 height	Y11 shoe size
1	F	145	2	154	3	160	4
2	F	153	3	160	5	167	5
3	F	137	1	147	3	154	4
4	F	141	2	156	4	165	6
5	F	149	2	160	3	170	4
6	M	131	3	157	5	165	7
7	M	151	5	159	7	175	9
8	M	141	3	155	5	171	8
9	M	150	4	165	5	180	10
10	M	129	3	135	4	155	7

1 **a** What was the height of pupil 1 in Year 7?

 b What was the height of pupil 6 in Year 7?

 c Find the total of the heights of the boys in Year 7.

 d Use your answer to **c** to find the mean height of the Year 7 boys.

 e Find the total of the heights of the girls in Year 7.

 f Use your answer to **e** to find the mean height of the Year 7 girls.

 g Who were the highest on average, the boys or the girls?

2 **a** Write down the smallest shoe size of the Year 7 boys.

 b Write down the biggest shoe size of the Year 7 boys.

 c Use your answers to **a** and **b** to find the **range** of the shoe sizes.

Exercise 11:8

1 a Find the mean height of the girls in Year 9.
   b Find the mean height of the boys in Year 9.

2 a Write down the smallest girl's height in Year 9.
   b Write down the biggest girl's height in Year 9.
   c Find the range of the girls' heights in Year 9.
   d Find the range of the boys' heights in Year 9.

3 a Look at the shoe sizes of the Year 9 boys.
      Write them down in order. Start with the smallest.
   b Which size is most common?
      This is the modal shoe size.
   c Make a list of the shoe sizes of the Year 9 girls.
   d What is the modal shoe size for the girls?

4 a Find the total of the Year 9 boys' shoe sizes.
   b Find the mean shoe size for the Year 9 boys.
● c Explain why this is not a sensible average to use for shoe size.

Exercise 11:9

● 1 a Find the mean height of the Year 11 girls.
      b Find the range of the heights of the Year 11 girls.
      c Find the mean height of the Year 11 boys.
      d Find the range of the heights of the Year 11 boys.
      e Find the modal shoe size of the Year 11 boys.
      f Find the modal shoe size of the Year 11 girls.

Year 11 are doing a GCSE statistics project. They are looking at how their heights and shoe sizes have changed since they came to school.

Exercise 11:10

1 This is a paragraph from Rubina's report. Copy the paragraph.
Fill in the gaps as you go.
Use your answers to Exercises 11:7, 11:8 and 11:9 to help you.

In Year 7 the girls' mean height is .... The boys' mean height is
.... This means that the .... are taller on average than the
....

The Year 7 girls' modal shoe size is .... and the boys' is
.... This means that the .... have bigger feet on average.

In Year 9, the .... are taller on average than the ....
The .... in Year 9 have larger feet on average than the ....

By Year 11, the boys' average height is .... and the girls' is
.... The girls' average shoe size is .... and the boys' is
.... This means that the .... are taller and have larger feet
on average than the ....

Exercise 11:11

W Growing up

You will need the data worksheet for this exercise.

The sheet contains information about thirty pupils.

Your task is to investigate how the heights and shoe sizes of the pupils have changed as they have grown up.

You should do whatever calculations you think are sensible.

You will also want to draw some diagrams.

When you have done all of your work you should write a report on what you have found out.

Write up your report neatly. Make sure that you include as much information as possible.

You can use the questions in this chapter to help you write your report.

You could do some work on heights and shoe sizes in your own class then compare it with the data given on the sheet.

1 These are the results of a girls' long jump competition.
The table shows the length of the jumps in centimetres.

Pupil	1st jump	2nd jump	3rd jump
Anne	258	260	310
Barbara	210	241	264
Christine	310	304	309
Davida	150	290	275
Evelyn	350	320	364
Firn	240	210	295

a How far did Barbara jump on her 1st jump?
b How far did Evelyn jump on her 3rd jump?
c Who did the best 1st jump?
d Who won the competition?
e Find the mean of Anne's jumps.
f Find the mean of Davida's jumps.
g Write down the median of Firn's jumps.
h Work out the range of Evelyn's jumps.

2 Howard works in a sports shop.
The shop sells football strips in three sizes.
The sizes are small, medium and large.

This bar-graph shows how many of each type Howard sold in one week.

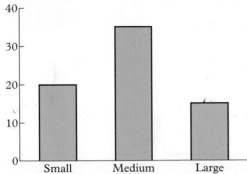

Bar-chart to show sales of football strips

a How many small strips did Howard sell?
b How many strips did Howard sell in the week?
c What was the modal size of strip sold?

3 Terry's Taxis advert says 'Average journey price £4'.
Mike's Minicabs says 'Average journey price £3.50'.

Here are some of the prices charged by the two firms.

Terry	£5	£1.90	£3.50	£4.50	£5	£2.10	£5.20	£4.80
Mike	£3.50	£8	£3.50	£9	£5	£3.50	£6	£7.50

a Work out the mean for Terry's Taxis.
b Work out the mean for Mike's Minicabs.
c Which firm has the lower mean price?
d What is the mode for Terry's Taxis?
e What is the mode for Mike's Minicabs?
f Which firm has the lower modal price?
g Which average has each firm used in its advert?

4 Quickslim's adverts says 'On average, people lose 6 pounds in the first week'.
Some users of Quickslim recorded how much weight they lost.
Here are their results to the nearest pound.

6	8	7	6	7	5	2	3	6
5	7	7	6	9	6	6	7	5

a What was the biggest amount of weight lost?
b What was the smallest amount of weight lost?
c Find the range of the weights lost.
d Find the mean amount of weight lost.
e Not everybody lost the average amount.
How does the range help you to show this?

5 Nadia owns 5 CDs.
She notices that they last for different times.
The lengths in minutes of the CDs are:

    55   64   69   45   63

a Work out the median length of the CDs.
b Work out the mean length of the CDs.

6 Nathan has a choice of two buses to get to town.
The buses are Bob's Buses and Carol's Coaches.

This is how long Nathan has waited the last five times he has used Bob's Buses:

   8 min   10 min   12 min   9 min   11 min

This is how long Nathan has waited the last five times he has used Carol's Coaches:

   2 min   16 min   5 min   15 min   7 min

a Find the mean waiting time for Bob's Buses.
b Find the mean waiting time for Carol's Coaches.
c Find the range for Bob's Buses.
d Find the range for Carol's Coaches.
e Use your answers to **a**–**d** to say which bus company you would choose.

7 Northfield school has a maths quiz team.
They are taking part in an important competition.
They have to choose between Ben and Anisha to take part.
This is how many points out of 10 Ben has scored in practice:

   7   8   6   8   7   6

This is how many points out of 10 Anisha has scored in practice:

   3   10   6   10   4   9

a Find the mean of Ben's scores.
b Find the mean of Anisha's scores.
c Find the range of Ben's scores.
d Find the range of Anisha's scores.
e Who would you choose for the team?
Explain your answer.

1 Joe asks his friends how much they spend on lunch in one week. These are his results:

£5	£6.50	£6.30	£3	£4.70
£6	£5.40	£2.90	£4.60	£6
£5.20	£3.40	£6.50	£4.30	£5.20
£6.20	£4.60	£5.30	£5.60	£7

a Find the total amount they spend.
b Find the mean amount they spend.
c Write the amounts in order. Start with the smallest.
d Write down the median amount.
e Copy this tally-table. Fill it in.

Amount	Tally	Total
£2.01–£3		
£3.01–£4		
£4.01–£5		
£5.01–£6		
£6.01–£7		

f Draw a bar chart to show the results.

2 These are the heights of 12 pupils in class 9W. The heights are in metres.

1.56	1.64	1.58	1.29
1.42	1.67	1.43	1.47
1.57	1.70	1.28	1.51

a Find the mean height.
b Find the median height.
c Find the range of the heights.
d Explain why you can't find the modal height.
e Two other pupils in the class have heights of 1.45 m and 1.63 m. Find the mean if these two people are included.

- **Mean**

  To find the **mean** of a set of data:
  a find the total of all the data values,
  b divide the total by the number of data values.

  Example

  Here are the weights of 6 sprinters in kilograms.
  Find their mean weight.

  The total is:

  $$88 + 90 + 79 + 94 + 86 + 91 = 528 \text{ kg}$$

  The mean is $528 \div 6 = 88$ kg

- **Mode**

  The **mode** is the most common or most popular data value.
  It is sometimes called the **modal value**.

  Example

  These are the winning distances in the men's javelin for the
  past few Olympics. They are given to the nearest metre.
  Find the modal distance.

  90   90   95   91   89   84   90

  There have been more **90** m winning throws than anything else.
  The mode is **90** m.

- **Median**

  To find the median you put all the data values in order of size.
  The **median** is then the middle value.

  Example

  The number of medals won by Japan at the Olympics from
  1972–1996 are:

  29   25   0   32   14   22   14

  Find the median number of medals Japan has won.

  Write the numbers in order. Start with the smallest.

  0   14   14   (22)   25   29   32

  Find the middle number. It is **22**.

  The median number of medals is **22**.

- **Range**

  The **range** of a set of data is the biggest value take away the
  smallest value.

1 These are the heights in centimetres of 5 sprinters.
   a Find the total of the heights.
   b Find their mean height.

       175   180   185   176   184

2 This table shows the medals won at the Olympic Games by France.

Year	Gold	Silver	Bronze	Total
1996	15	7	14	36
1992	8	5	16	29
1988	6	3	6	15

   a How many gold medals did France win in 1992?
   b How many bronze medals did they win in 1988?
   c How many bronze medals have they won altogether in the last three Olympics?
   d Find the mean number of bronze medals won in the last three Olympics.
   e Find the mean number of silver medals won in the last three Olympics.

3 Here are the number of Olympic medals won by New Zealand from 1972–1996.

       3   4   0   11   13   10   6

   a Write these numbers in order. Start with the smallest.
   b Find the median number of medals won by New Zealand.

4 Six plumbers were asked to give a price for doing a job.
The prices were:

       £65   £50   £85   £110   £70   £85

   a Find the mean of these prices.
   b Find the range of these prices.
   c Find the median price.

12 Try angles

These satellite dishes are set at a precise angle of elevation and direction (east) to receive TV broadcasts from a satellite orbiting the Earth. Find out what the angle is.

1   Names for angles

The pilot knows that when the wings of the aeroplane on the dial are resting on the horizontal line the plane is flying level. If there is an angle between the wings and the horizontal line then the aeroplane is turning.

The aeroplane may be turning clockwise or anticlockwise.

Example

An aeroplane is flying north.
In which direction will it be flying after making these turns?
a Half a turn anticlockwise   **b** A quarter turn clockwise

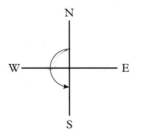   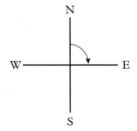

Answer: south   east

Exercise 12:1

1 An aeroplane is flying north.
In which direction will it be flying after making these turns?
a A quarter turn anticlockwise.
b Half a turn clockwise.

2 An aeroplane is flying south.
   In which direction will it be flying after making these turns?
   a A quarter turn clockwise.
   b Half a turn clockwise.
   c A quarter turn anticlockwise.

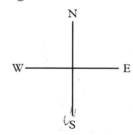

An angle is how we measure turn.
Angles are measured in degrees.

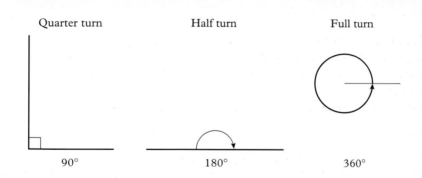

Quarter turn	Half turn	Full turn
90°	180°	360°

Acute angle   Any angle less than
          90° is called an
          acute angle.

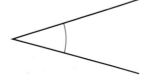

Obtuse angle   Any angle between
          90° and 180°
          is called an
          obtuse angle.

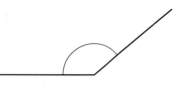

Reflex angle   Any angle bigger than
          180° is called a
          reflex angle.

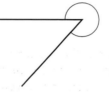

3 Write down the name of each of these angles.
Choose from: acute, right, obtuse, reflex.

a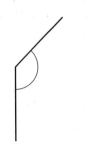

c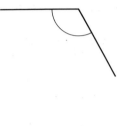

e

b

d

f

4 Laura has to pick the acute angles from this list.
Which ones should she pick?
45°   253°   28°   310°   125°   90°   95°   80°

5 Richard had to pick out reflex angles from another list.
These are the ones he chose.
135°   260°   195°   210°   185°   179°   345°
Write down the ones that are correct.

6 Which of these angles are obtuse?
55°   110°   85°   175°   185°   40°   125°   99°

7 Write down the name of the angles marked with letters.

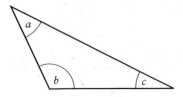

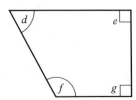

We use a protractor or angle measurer to measure angles.
They have two scales; a clockwise scale and an anticlockwise scale.

Example

Use the protractor to measure ∠ABC.
Remember: ∠ABC is read as angle ABC.

Put the cross of the protractor on the point of the angle.
The zero line of the protractor must lie on a side of the angle.

Look for the scale that starts at zero on the line.
Use this scale. Angle ABC is 52°.

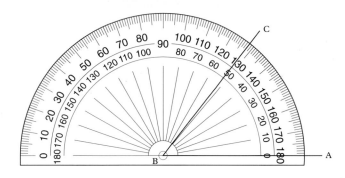

Exercise 12:2

1 Use the diagram to find these angles.

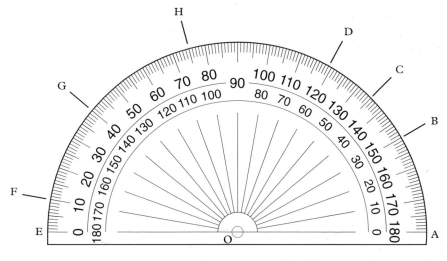

a ∠AOB	**c** ∠AOD	**e** ∠EOG
b ∠AOC	**d** ∠EOF	**f** ∠EOH

2 Measure these angles.
  Use a protractor or angle measurer.

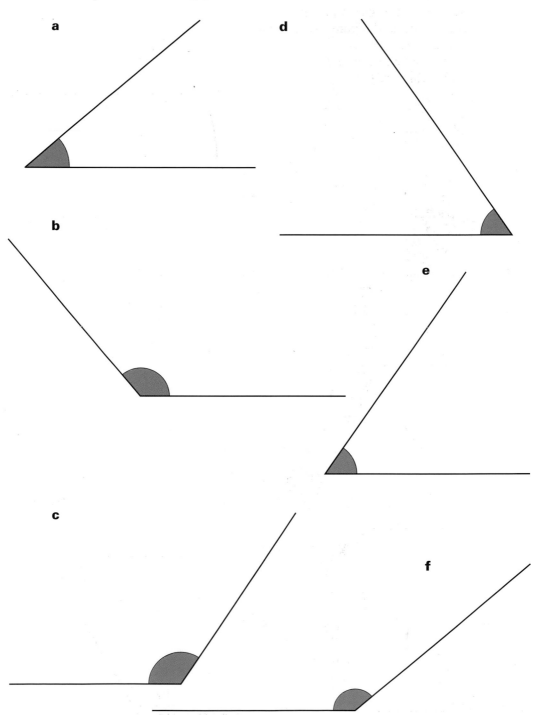

a

d

b

e

c

f

3 **a** Measure angle BAC.
    b Measure angle DAC.
    c Add the two angles together.
    d Draw another diagram like this one.
        Use different angles.
        Measure the two angles.
        Add them together.

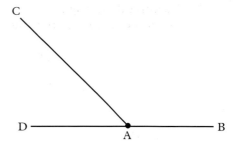

Angles on a straight line	**Angles on a straight line** add up to 180°.

4 Find the pairs of angles that will fit together on a straight line.

a

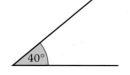

40°

d

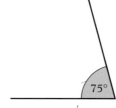

75°

b

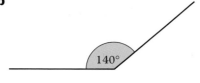

140°

e

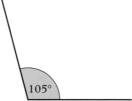

105°

c

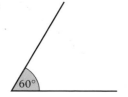

60°

f

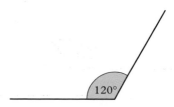

120°

5 **a** Draw a triangle.
Make each side at least 6 cm long.
b Measure each angle of the triangle.
Label each angle with its size.
c Add the three angles together.
Write down the answer inside your triangle.
d Draw two more triangles.
Measure their angles.
Add the three angles together.
e What do you notice about the answers to
parts **c** and **d**?

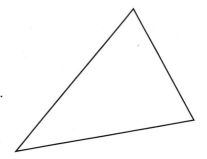

Angles of a triangle	The **angles of a triangle** add up to 180°.

6 Three of these angles will fit together to make a triangle. Find the
three angles.
    a 40°   **b** 50°   **c** 60°   **d** 80°

7 Harry has measured the angles of each of these triangles.
He has got some angles wrong.
Which triangles are wrong? Do not use a protractor.

a

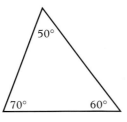

c

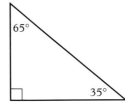

b

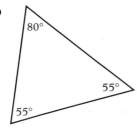

d

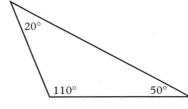

Example   Draw an angle of 40° at **A**.

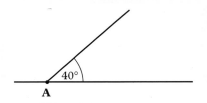

Put the protractor on the
line with the cross point
at **A**.

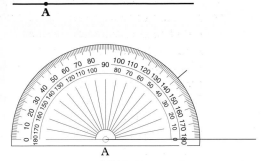

Find 40° on the scale.
Make a mark at the 40°
line of the protractor.

Remove the protractor.

Join **A** to your mark
with a straight line.

Label the angle 40°.

Exercise 12:3

1 Draw these angles.
Start with a line like
this each time.
Mark a point **A** on each line.

    a 60°   **c** 35°   **e** 75°

    b 80°   **d** 55°   **f** 20°

2 Draw these angles.

    a 42°   **c** 28°   **e** 73°

    b 59°   **d** 84°   **f** 36°

3 Draw these angles.

    a 120°   **c** 100°   **e** 155°

    b 160°   **d** 135°   **f** 95°

2   Drawing triangles

William wants to cut out a right angled triangle.
He has made a sketch.
He needs to draw the outline on the wood.
William can then cut along the outline.

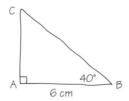

Example

Make an accurate drawing of William's triangle.

Draw a line AB
6 cm long.

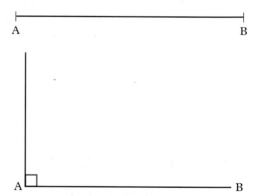

Draw an angle
of **90°** at A.

Draw an angle
of **40°** at B.

Label the point
C where your two
lines cross.

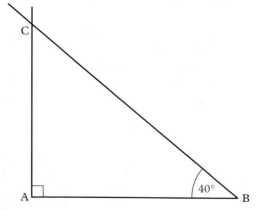

Exercise 12:4

1 William sketched some other triangles.
Make an accurate drawing of each one.

a
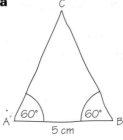

b
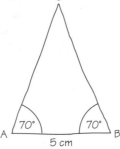

c
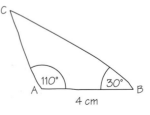

2 Rhian is making a clock.
This is her design.
She is using a triangular piece of wood.
Make an accurate drawing of the piece of wood.

3 Jenny is also using triangles in
her design.
She is making a letter rack.
Jenny needs two triangles of
wood.
Here are her sketches of the
two triangles.
Make an accurate drawing
of each one.

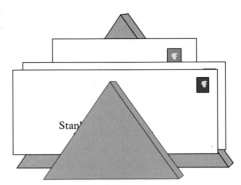

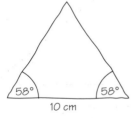

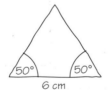

4   Rob is making Christmas decorations.
     He is making trees out of wood.
     This is a rough sketch of his design.
     Make an accurate drawing of the
     triangular part of the tree.

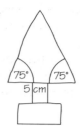

5   Marie is drawing a net of a pyramid.
     This is a sketch of her net.
     Make an accurate drawing of one of
     the triangles.

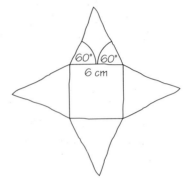

6   Lisa is making this sign to direct
     people to the hall.
     Make an accurate drawing of the
     triangle in the sign.

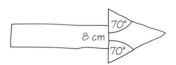

7   Peter is making a badge.
     This is his design.
     Make an accurate drawing of the badge.
     Draw the rectangle first.
     Then draw the triangle.

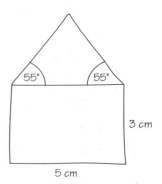

1 An aeroplane is flying east.
In which direction will it be flying after making these turns?
  a A quarter turn clockwise.
  b Half a turn clockwise.
  c A quarter turn anticlockwise.

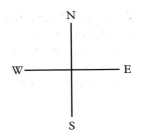

2 An aeroplane is flying west.
In which direction will it be flying after making these turns?
  a A quarter turn clockwise.
  b Half a turn anticlockwise.
  c A quarter turn anticlockwise.
  d Three quarters of a turn clockwise.

3 Look at these angles:
  125° 27° 350° 190° 95° 30° 147° 225°
  a Which are acute angles?
  b Which are obtuse angles?
  c Which are reflex angles?

4 Use the diagram to find these angles.
  a AOC   **d** BOD   **g** BOA
  b AOE   **e** AOG   **h** AOD
  c BOC   **f** BOH   **i** BOE

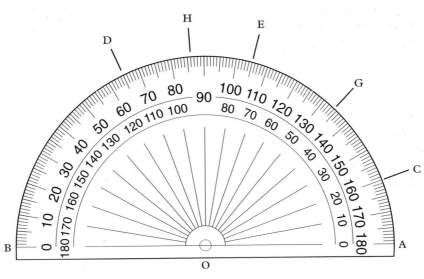

5 Find the pairs of angles that will fit together on a straight line.

a

25°

d

150°

g

80°

b

155°

e

100°

h

65°

c

60°

f

115°

i

30°

6 Find the missing angle.

a

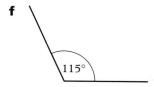

? 140°

c

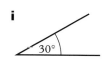

? 35°

b

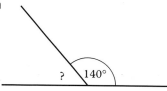

?

d

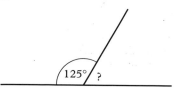

125° ?

7 Which triangle is impossible? Show your working.

a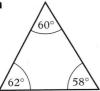

b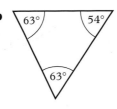

c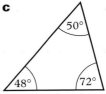

8 Petra is making a wooden stand for her
diary and address books.
She needs two rectangles and two triangles.
This is Petra's sketch of the triangular piece
of wood.
Make an accurate drawing of the triangle.

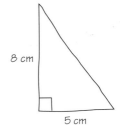

9 Make an accurate drawing of each of these triangles.

a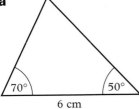

b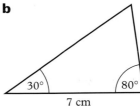

10 Make an accurate drawing of each of these triangles.

a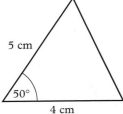

b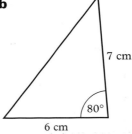

1 An aeroplane is flying southwest.
In which direction will it be flying
after making these turns?
  a A quarter turn clockwise.
  b Half a turn anticlockwise.
  c A quarter turn anticlockwise.
  d Three quarters of a turn clockwise.

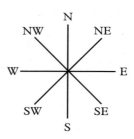

2 An aeroplane is flying northeast.
In which direction will it be flying after turning through these angles?
  a 90° clockwise
  b 45° anticlockwise
  c 135° anticlockwise
  d 225° clockwise

3 Find the missing angles.

a

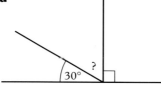

d

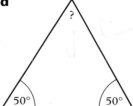

b

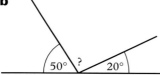

e

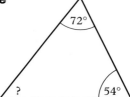

c

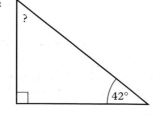

f

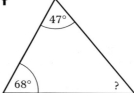

- **Acute angle**   Any angle less than 90° is called an **acute angle.**

- **Obtuse angle**   Any angle between 90° and 180° is called an **obtuse angle.**

- **Reflex angle**   Any angle bigger than 180° is called a **reflex angle.**

- **Angles on a straight line**   **Angles on a straight line** add up to 180°.

- **Angles of a triangle**   The **angles of a triangle** add up to 180°.

- *Example*   Make an accurate drawing of this triangle.

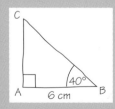

(1) Draw a line AB 6 cm long.

(2) Draw an angle of 90° at A.

(3) Draw an angle of 40° at B.

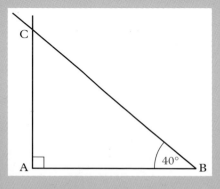

(4) Label the point C where your two lines cross.

1 A helicopter is flying west.
In which direction will it be flying after making these turns?
a A quarter turn anticlockwise.
b Half a turn clockwise.
c Three quarter turn clockwise.

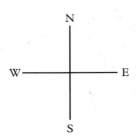

2 Look at these angles.
    225°   62°   150°   290°   85°   140°   27°   305°
a Which are acute angles?
b Which are obtuse angles?
c Which are reflex angles?

3 Why are these diagrams impossible?
Explain your answer.

a
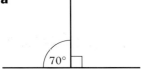

c
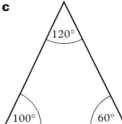

b
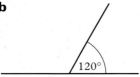

d
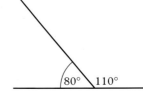

4 This is a sketch of a triangle.
Make an accurate drawing of the triangle.

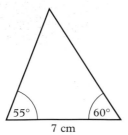

13 Formulas

For a moving object

$E = mc^2$

where E = energy,
c is the velocity of
light
and m is mass

In simple terms
this states that
all energy has
mass. It was
devised by the
German-born
scientist Albert
Einstein.

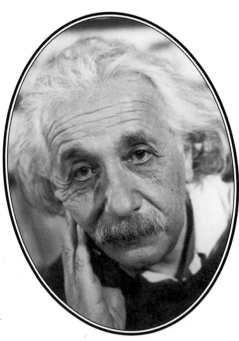

1 Using letters

· ·

... *continued* on page 259 ...

You can write addition sums as multiplications.

Example $3 + 3 + 3 + 3 = 4 \times 3 = 12$

Exercise 13:1

Copy these and fill them in.

1 $5 + 5 + 5 = 3 \times 5 = \ldots$

2 $2 + 2 + 2 + 2 = \ldots \times 2 = \ldots$

3 $3 + 3 + 3 + 3 + 3 = \ldots \times 3 = \ldots$

4 $10 + 10 + 10 + 10 + 10 + 10 = \ldots \times 10 = \ldots$

5 $7 + 7 + 7 = \ldots \times 7 = \ldots$

In algebra you can use letters instead of numbers.

Example

$m + m + m + m = 4 \times m$

In algebra you miss out the $\times$ sign.

You write $4 \times m$ as $4m$.

Write these in a short form using algebra.

6 $a + a + a = ... \times a$

       $= 3a$

7 $p + p + p + p = ... \times p$

            $= 4p$

8 $h + h + h + h + h = ... \times h$

                $= ...$

9 $t + t = ... \times t$

       $= ...$

10 $j + j + j + j + j + j + j = ... \times j$

                 $= ...$

11 $d + d + d + d + d = ... \times d$

               $= ...$

12 $k + k + k = ... \times k$

         $= ...$

13 $w + w + w + w + w + w = ... \times w$

                   $= ...$

Karen has 3 bags of counters.
Each bag has **c** counters inside.
There are $3 \times$ **c** counters in the bags.
Karen also has 2 extra counters.

Karen has **3c + 2** counters altogether.

You cannot add letters to numbers.
You cannot add $3c$ to 2.

Exercise 13:2

Write down the total number of counters.

1

2

3

5

4

6

- -

Example

Steve has 2 bags of marbles.
Each bag has **m** marbles inside.
a How many marbles does
Steve have altogether?
b Steve plays a game and loses
4 marbles.
How many marbles does he have now?

a Steve starts with $m + m = 2m$ marbles.
b After the game Steve has $2m - 4$ marbles.

Exercise 13:3

For each of these children write down
a the number of marbles they have at the start of the game,
b the number of marbles they have after the game.

1

Start	During the game	After the game
Liz 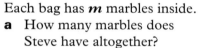	lost 3 marbles	
a ... m		**b** 2m − ...

256

	Start	During the game	After the game
2	Pranav	lost 5 marbles	
	a ... m		**b** ... m − ...
3	Tracy	won 4 marbles	
	a m		**b** m ... 4
4	Sherene	won 2 marbles	
	a		**b**
5	Don	lost 6 marbles	
	a		**b**
6	Andy	won 8 and lost 2 marbles	
	a		**b**

Perimeter   The total distance around the outside of a shape is its
perimeter.

Examples   Write down the perimeter p of each of these shapes.

 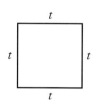 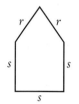

$p = 3 + 3 + 3 + 3$ $p = t + t + t + t$ $p = s + s + s + r + r$ $p = d + d + 3$
$p = 12$ $p = 4t$ $p = 3s + 2r$ $p = 2d + 3$

Exercise 13:4

Write an expression for the perimeter p of each of these shapes.

1

$p = \ldots a$

4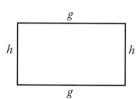

$p = \ldots g + \ldots h$

7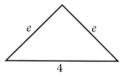

2

5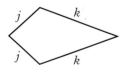

8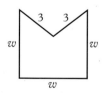

3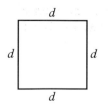

6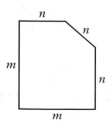

● **9**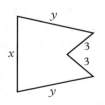

2 Patterns

... continued on page 265 ...

Exercise 13:5

1 Alan is making patterns with red and blue counters.

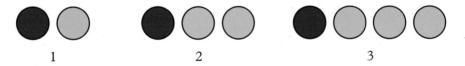

      1   2   3

   a Draw the next two patterns.
   b How many more blue counters does Alan add each time he makes
       a new pattern?
   c How many counters will there be altogether in pattern number 6?
   d How many counters will there be altogether in pattern number 10?
   e Alan's rule for the pattern is $n + 1$.
       What does the 1 stand for in the rule?
   f What does the n stand for in the rule $n + 1$?
   g Alan makes a pattern using 15 counters altogether.
       How many blue counters does he use?

2   Sandra is making patterns with red and yellow tiles.

1

2

3

a   Draw the next two patterns.

b   How many more yellow tiles does Sandra add each time she makes a new pattern?

c   How many tiles will there be altogether in pattern number 6?

d   How many tiles will there be altogether in pattern number 7?

e   How many tiles will there be altogether in pattern number 12?

f   Sandra's rule for the pattern is $n + 2$.
What does the 2 stand for in the rule?

g   What does the n stand for in the rule $n + 2$?

h   Sandra makes a pattern using 20 tiles altogether.
How many red tiles does she use?

i   Sandra makes a pattern using 20 tiles altogether.
How many yellow tiles does she use?

3 Mark is making these patterns using red and yellow tiles.

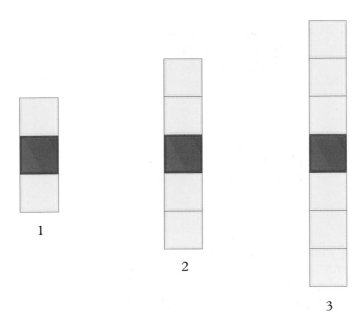

1

2

3

a How many more yellow tiles does Mark add each time he makes a new pattern?

b How many tiles will there be altogether in pattern number 4?

c How many tiles will there be altogether in pattern number 10?

Mark's rule for the pattern is $2n + 1$.

d What does the 1 stand for in the $2n + 1$ rule?

e What does the $2n$ stand for in the $2n + 1$ rule?

f Mark wants to make pattern number 8.
How many yellow tiles does he need?

g Mark has 5 red tiles and 24 yellow tiles.
What is the number of the biggest pattern that Mark can make?

h Mark makes a pattern using 21 tiles altogether.
How many red tiles does he use?

i Mark makes a pattern using 41 tiles altogether.
How many yellow tiles does he use?

4   Jane is making patterns using green and yellow tiles.

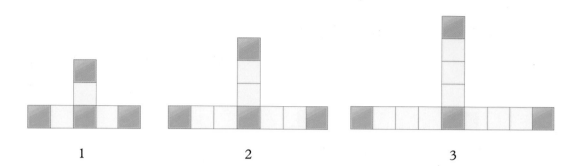

1   2   3

a   How many more yellow tiles does Jane add each time she makes a new pattern?

b   How many tiles will there be altogether in pattern number 4?

c   How many tiles will there be altogether in pattern number 8?

Jane's rule for the pattern is $3n + 4$.

d   What does the 4 stand for in the $3n + 4$ rule?

e   What does the $3n$ stand for in the $3n + 4$ rule?

f   Jane wants to make pattern number 11.
How many yellow tiles does she need?

g   How many tiles does Jane need altogether to make pattern number 11?

h   Jane makes a pattern using 34 tiles altogether.
How many yellow tiles does she use?

i   Jane has 8 green tiles and 36 yellow tiles.
What is the number of the biggest pattern that Jane can make?

Exercise 13:6

1 These patterns are made from beads and matchsticks.

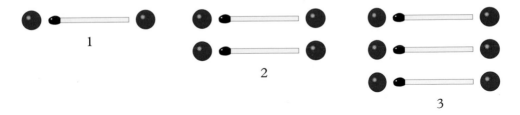

1

2

3

a Draw the next two patterns.
b Copy this table and fill it in.

Number of matchsticks	1	2	3	4	5
Number of beads	2	4	...	...	...

+?   +?   +?   +?

c How many beads do you add each time?
d Copy this and fill it in:
    number of beads = ... × number of matchsticks

2 These patterns are made from yellow and blue counters.

1   2   3

a Draw the next two patterns.
b Copy this table and fill it in.

Number of blue counters	1	2	3	4	5
Number of yellow counters	4	8	...	...	...

+?   +?   +?   +?

c How many yellow counters do you add each time?
d Copy this and fill it in:
    number of yellow counters = ... × number of blue counters

3 These patterns are made from yellow and blue counters.

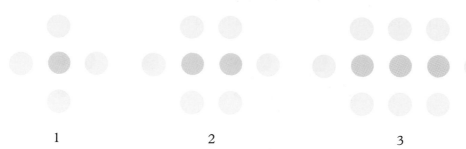

1   2   3

a Draw the next two patterns.
b Copy this table and fill it in.

Number of blue counters	1	2	3	4	5
Number of yellow counters	4	6	...	...	...

c How many yellow counters do you add each time?

The first part of the formula is:
number of yellow counters = 2 × number of blue counters + ?

d Copy this table.
Finish the row of green numbers.
Use the first part of the formula.

Number of blue counters	1	2	3	4	5
	2 +?	4 +?	+?	+?	+?
Number of yellow counters	4	6	...	...	...

e What do you add to the green numbers to get the number of yellow counters?

f Copy this and fill in the rule for the number of yellow counters.
number of yellow counters = ... × number of blue counters + ...

3   Using formulas

...

Here are some algebra puzzles for you to try.

Exercise 13:7

1   The numbers in the end squares add up to the numbers in the middle square like this:

Copy these.

a   Fill in the missing numbers.

b   Write down the value of each letter.

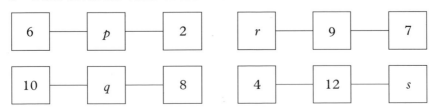

2 To get the number in the middle square take *b* away from *a* like this:

Copy these.

a Fill in the missing numbers.

b Write down the value of each letter.

3 The numbers in ◯ and ◯ are multiplied to get the number in ☐ like this:

Copy these.

a Fill in the missing numbers.

b Write down the value of each letter.

4 Sue is joining numbers in a circle.
She is using the rule
'double the number then add 1'.
Copy Sue's list of joins.
Fill it in.
1 joins to 3
2 joins to …
3 joins to …
… joins to 11

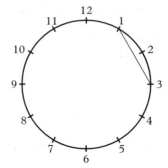

A function machine
can have two steps:

→ | × 3 | → | + 2 | →

Example

10 → | × 3 | →30→ | + 2 | → 32

The inverse machine is:

← | ÷ 3 | ← | − 2 | ←

Example

4 ← | ÷ 3 | ←12← | − 2 | ← 14

Exercise 13:8

1 Rob put some numbers into this function machine:

→ | × 10 | → | + 3 | →

Copy these function machines.
Fill in the spaces.

3 → | × 10 | →…→ | + 3 | → 33

5 → | × 10 | →…→ | + 3 | → …

7 → | × 10 | →…→ | + 3 | → …

10 → | × 10 | →…→ | + 3 | → …

2 Jackie put some numbers into this function machine.

a Copy these function machines.
Fill in the spaces.

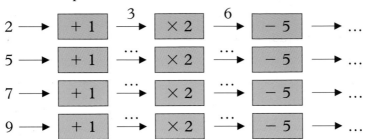

Jackie draws the inverse function machine:

b Copy the inverse function machine.
Fill it in.
Use it to find the inverse of 25.

3 David has got two special function machines.
The pink function machine is the inverse of the blue function machine.

Remember: the square of $3 = 3^2 = 3 \times 3 = 9$.

a Copy the blue function machines.
Fill in the spaces.

$1 \longrightarrow$ work out the square of $\longrightarrow 1$

$2 \longrightarrow$ work out the square of $\longrightarrow 4$

$3 \longrightarrow$ work out the square of $\longrightarrow$ ...

$4 \longrightarrow$ work out the square of $\longrightarrow$ ...

b Copy the pink function machines.
Fill in the spaces.

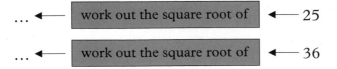

Exercise 13.9

Investigation: Surrounded

You need some squared paper.

1 Colour four squares in red like this.
Put them near the middle of a piece of
squared paper.

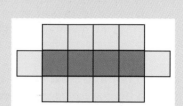

2 **a** Look at the yellow squares.
They *surround* the red squares.
Colour the squares which touch
your red squares.
b How many yellow squares did
you need?

The yellow squares are surround number 1.

3 **a** Look at the green squares.
They *surround* the yellow squares.
Colour the squares which touch
your yellow squares.
b How many green squares did you
need?

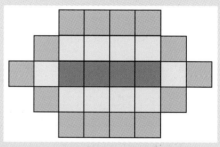

The green squares are surround number 2.

4 You can surround the green squares with blue squares.
The blue squares will be surround number 3.
How many blue squares will you need?

5 How many squares will you need for surround number 4?

6 How many squares will you need for surround number 20?

7 Can you write down any rules?

8 You could try a different number of squares to start with.

9 You could try a different shape to start with.

10 Write a report about what you have found.

1 Write these in a short form using algebra:
 a $a + a + a + a = \ldots \times a = 4a$
 b $t + t + t + t + t = \ldots \times t = \ldots$
 c $f + f + f + f + f + f + f = \ldots \times f = \ldots$

2 Ned has 4 bags of marbles.
 Each bag has m marbles inside.
 a How many marbles has Ned got altogether?
 b Ned loses 4 marbles in a game.
   How many marbles has Ned got left?

3 3 minibuses set off on a trip.
 There are p people in each bus.
 a How many people are there altogether?
 b 6 people do not come back on the minibuses.
   How many people do come back on the minibuses?

4 Write an expression for the perimeter p of each of these shapes.

 a **b** 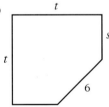 **c**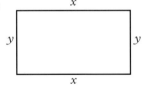

5 Michelle is making patterns using green and yellow tiles.

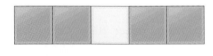

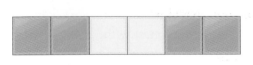

a How many more yellow tiles does Michelle add each time she makes a new pattern?

b How many tiles will there be altogether in pattern number 4?

c How many tiles will there be altogether in pattern number 8?

Michelle's rule for the pattern is $n + 4$.

d What does the 4 stand for in the $n + 4$ rule?

e What does the n stand for in the $n + 4$ rule?

f Michelle wants to make pattern number 11.
How many yellow tiles does she need?

g How many tiles does Michelle need to make pattern number 11?

h Michelle makes a pattern using 21 tiles altogether.
How many yellow tiles does she use?

i Michelle has 4 green tiles and 21 yellow tiles.
What is the number of the biggest pattern that Michelle can make?

6 Matt put some numbers into this function machine:

IN ⟶ ┃ + 4 ┃ ⟶ ┃ × 5 ┃ ⟶ ┃ − 7 ┃ ⟶ OUT

Copy this list.
Fill in the spaces.

IN   OUT
  2 ⟶ 6 ⟶ 30 ⟶ 23

  5 ⟶ ... ⟶ ... ⟶ ...

  7 ⟶ ... ⟶ ... ⟶ ...

  9 ⟶ ... ⟶ ... ⟶ ...

7 Rachel put a number into this function machine:

⟶ ┃ × 2 ┃ ⟶ ┃ − 6 ┃ ⟶

She gets the number 4 out.
What number did she put in?

1   Some Year 9 pupils are going to Alton Towers.
The total cost (in £) is worked out using the formula:
cost = 11 × number of *pupils* + 100
or **c = 11p + 100**
  a   How much would it cost to take 50 pupils?
  b   How much would it cost to take 100 pupils?
  c   The cost of a trip is £760.
       How many pupils go on the trip?

2   The square in the middle is worked out using the formula $2a + b$.
  a   Copy each of the patterns.
  b   Fill in the middle numbers.

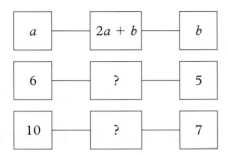

3   Lindsey is doing an investigation into patterns.
She finds a rule.
Her rule is $t = 2n + 1$.
She decides to draw a graph of her rule.
  a   Copy this table.

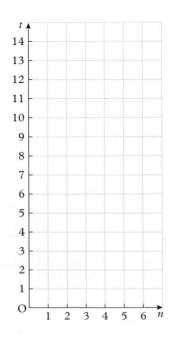

n	$2n + 1$	t
1	$2 \times 1 + 1$	3
2	$2 \times 2 + 1$	5
3		
4		
5		
6		

  b   Fill in the rest of the table.
  c   Copy the axes shown.
  d   Plot the rest of the points.
  e   Join the points with a straight line.
  f   Label your line with the rule $t = 2n + 1$

- In algebra you can use letters instead of numbers.

  Examples **1** $m + m + m + m = 4 \times m$

  In algebra you miss out the $\times$ sign.
  You write $4 \times m$ as $4m$.

  2 Steve has 2 bags of marbles.
  Each bag has m marbles inside.

  a How many marbles does
  Steve have altogether?

  b Steve plays a game and loses
  4 marbles.
  How many marbles does he have now?

  a Steve starts with $m + m = 2m$ marbles.

  b After the game Steve has $2m - 4$ marbles.

- **Perimeter**   The total distance around the outside of a shape is its
  perimeter.

  Examples   Write an expression for the perimeter p of each of these shapes.

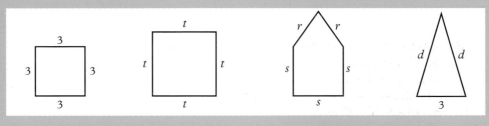

$p = 3 + 3 + 3 + 3$ $p = t + t + t + t$ $p = s + s + s + r + r$ $p = d + d + 3$
$p = 12$ $p = 4t$ $p = 3s + 2r$ $p = 2d + 3$

- A function machine
  can have two steps:

  Example

  The inverse machine is:

  Example

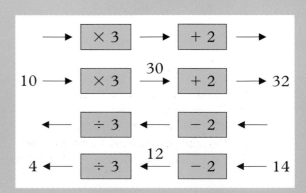

1 Write these in a short form using algebra:

    a $a + a + a = \ldots \times a = \ldots$ **b** $g + g + g + g + g = \ldots \times g = \ldots$

2 Ben has 3 bags of maltesers. Each bag has m maltesers inside.

    a How many maltesers has Ben got altogether?

    b Ben eats 12 maltesers, how many maltesers has he got left?

3 Write an expression for the perimeter p of this shape.

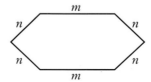

4 Jenny is making patterns using black and white counters.

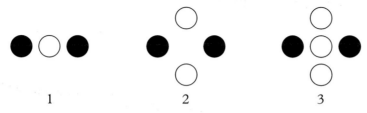

          1   2   3

    a Draw the next two patterns.

        Jenny's rule for her patterns is $n + 2$.

    b What does the 2 stand for in the $n + 2$ rule?

    c What does the n stand for in the $n + 2$ rule?

    d Jenny has made a pattern using 25 counters altogether. How many of the counters are white?

5 Ryan put some numbers into this function machine:

IN $\longrightarrow$ $\boxed{+\ 6}$ $\longrightarrow$ $\boxed{\times\ 3}$ $\longrightarrow$ $\boxed{-\ 5}$ $\longrightarrow$ OUT

Copy this list.
Fill in the spaces.

IN   OUT

2 $\longrightarrow$ 8 $\longrightarrow$ 24 $\longrightarrow$ 19

4 $\longrightarrow$ … $\longrightarrow$ … $\longrightarrow$ …

6 $\longrightarrow$ … $\longrightarrow$ … $\longrightarrow$ …

10 $\longrightarrow$ … $\longrightarrow$ … $\longrightarrow$ …

14 Fractions

Fold a sheet of paper in half, and in half again, and again as many times as you can.

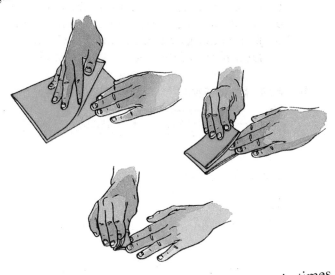

You will not be able to fold it more than six times.

What fraction of the whole sheet is made at each fold?

1 Fractions

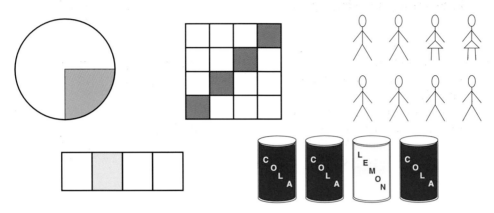

Spot the quarter, $\frac{1}{4}$, in each of these pictures.

Exercise 14:1

You need some sheets of plain paper for this exercise.

1 a Fold a sheet of paper in **half**.
Unfold your paper.
b How many parts is your paper divided into?
c How many halves make a whole?
d Copy this and fill it in:

One half $= \dfrac{\ldots}{2}$

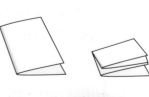

2 a Fold a sheet of paper in half.
Fold it in half again.
Unfold your paper.
b How many parts is your paper divided into?
c How many quarters make a whole?
d Write one quarter using figures.
e How many quarters of this sheet are shaded?
f Copy this and fill it in.

$\ldots$ quarters $= \dfrac{3}{\ldots}$

3 Fold your piece of paper again
so it looks like this.
   a How many parts is your paper divided into?
   b How many eighths make a whole?
   c Write one eighth using figures.

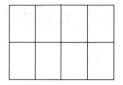

4 Look at the diagram.
   a How many eighths are blue?
   b What fraction is blue?
      Write your answer using figures.
   c How many eighths are red?
   d What fraction is red?
      Write your answer using figures.

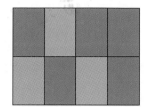

5 Start with a new piece of paper.
   a Fold it into three equal parts like this.
   b How many thirds make a whole?
   c Write one third using figures.
   d Write two thirds using figures.

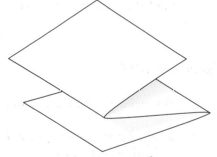

Sometimes it is difficult to fold the paper to show fractions.
You can use squares instead.

6 **a** How many fifths make a whole?
   b Write one fifth in figures.
   c Copy the diagram.
      Shade three fifths red.

     Shade $\frac{2}{5}$ green.

7 **a** How many ninths make a whole?
   b How many tenths make a whole?
   c Copy these and fill in the numbers.

$$1 = \frac{2}{2} = \frac{\ldots}{3} = \frac{4}{\ldots} = \frac{\ldots}{5} = \frac{6}{\ldots} = \frac{7}{7}$$

8 **a** How many halves make a whole?
   b How many halves make two wholes?
   c How many halves make three wholes?

9 **a** How many thirds make a whole?
   b How many thirds make two wholes?
   c How many thirds make three wholes?

10 You can see half of this ruler.
How long is the whole ruler?

11 You can see half of this exercise book.
How long is the whole book?

12 You can see one fifth of this window.
How many squares of glass are in
the whole window?

13

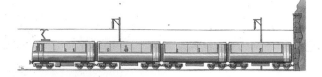

You can see two thirds of this train.
The rest of the train is in the tunnel.
How many carriages are there in the whole train?

Look at the fraction $\frac{5}{4}$ (five quarters).

The top number is larger than the bottom number.
The fraction is 'top heavy'.

Look at the diagram.
Five quarters make one whole and one quarter.

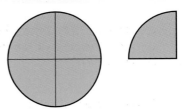

$$\frac{5}{4} = 1\frac{1}{4}$$

Exercise 14:2

Write each of these pictures
a as a 'top heavy fraction',
b as a whole number and a fraction.

1 $\dfrac{\ldots}{3} = 1\,\dfrac{1}{\ldots}$

5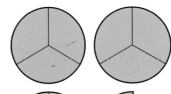

2 $\dfrac{\ldots}{4} = 1\,\dfrac{\ldots}{4}$

3

6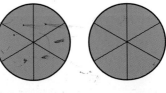

4

Example Mary has to find $\frac{1}{4}$ of 8.

She uses 8 counters to help her.
She folds her paper into four.
She shares the 8 counters
between the four parts.

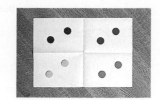

There are **2** counters on each part.
$\frac{1}{4}$ of 8 is **2**

Exercise 14:3

1 Look at the picture of Mary's paper.

   a How many counters are in three parts?

   b What is $\frac{3}{4}$ of 8?

2 You need a piece of paper folded like Mary's.
You also need 12 counters.

   a Share the counters between the four parts of your paper.

   b How many counters are in each part?

   c Copy and fill in:

      $\frac{1}{4}$ of 12 = ... $\frac{3}{4}$ of 12 = ...

3 Use the piece of paper from question **2** and 20 counters.
Find each of these.

   a $\frac{1}{4}$ of 20 = ...   **b** $\frac{3}{4}$ of 20 = ...

4 Fold your paper again so it looks like this.
You need 16 counters.
Use your paper to find

   a $\frac{1}{8}$ of 16   **c** $\frac{3}{8}$ of 16

   b $\frac{5}{8}$ of 16   **d** $\frac{7}{8}$ of 16

5 Fold a sheet of paper into 3 like this.

Use the paper and six counters to find:

a $\frac{1}{3}$ of 6 **b** $\frac{2}{3}$ of 6

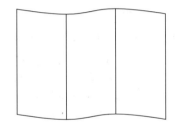

6 **a** How many counters do you need to find $\frac{1}{3}$ of 12?

b Use your counters to find $\frac{1}{3}$ of 12.

c Use your counters to find $\frac{2}{3}$ of 12.

7 Use squares and counters to find

a $\frac{1}{3}$ of 15 **b** $\frac{2}{3}$ of 15

Use squares and counters to find each of these.

8 **a** $\frac{1}{3}$ of 21 **b** $\frac{2}{3}$ of 21

9 **a** $\frac{1}{2}$ of 14 **b** $\frac{1}{2}$ of 16

10 **a** $\frac{1}{4}$ of 16 **b** $\frac{3}{4}$ of 16

11 **a** $\frac{1}{4}$ of 24 **b** $\frac{3}{4}$ of 24

· ·

Example Phil has no counters.

He needs to find $\frac{1}{5}$ of 10.

Phil knows he needs to divide 10 into five groups.

To find $\frac{1}{5}$ you divide by 5:

$10 \div 5 = 2$ $\quad$ $\frac{1}{5}$ of 10 is 2

Exercise 14:4

1 Write down the number you divide by to find these fractions of a number:

   a $\frac{1}{2}$ **b** $\frac{1}{3}$ **c** $\frac{1}{4}$ **d** $\frac{1}{6}$

2 Find $\frac{1}{2}$ of 8.

3 Find: **a** $\frac{1}{3}$ of 6   **b** $\frac{2}{3}$ of 6

4 Find: **a** $\frac{1}{4}$ of 40   **b** $\frac{3}{4}$ of 40

5 Find: **a** $\frac{1}{5}$ of 20   **b** $\frac{2}{5}$ of 20   **c** $\frac{3}{5}$ of 20   **d** $\frac{4}{5}$ of 20

How do you divide **seven** cakes between **four** children?
The children all love cakes. Each wants an equal share.

You have to work out $7 \div 4$. This can be written $\frac{7}{4}$.

The children each get **one whole** cake.
The remaining **three** cakes are cut into **four** quarters.

$7 \div 4$ can be written $\frac{7}{4} = 1\frac{3}{4}$ Each child gets $1\frac{3}{4}$ cakes.

Exercise 14:5

1 Share three apples equally between the two children.

2 Share five chocolate bars between three children.

3 Share five cakes between four children.

4 Share two oranges between three children.

5 Share four pizzas between five children.

Game: Fractions of 12

Here is a game for two people.

You need a blank dice.

You could stick plain paper on an ordinary dice.
You could use a spinner divided into six sections.

Write one of these fractions on each face of the dice.

$$\frac{1}{2} \quad \frac{1}{3} \quad \frac{1}{4} \quad \frac{1}{6} \quad \frac{1}{12} \quad \frac{2}{3}$$

Playing the game

Shahnaz and Jason are playing the fraction game.

Shahnaz rolls the dice.

She gets a $\frac{1}{4}$.

Shahnaz has to **find $\frac{1}{4}$ of 12**.

This is her score.

Now it is Jason's turn.

He gets $\frac{1}{6}$.

Jason has to find $\frac{1}{6}$ of 12.

This is his score.

Shahnaz and Jason keep playing.

The first person whose score adds up to 20 or more is the winner.

Play the game with a friend.

2 Equivalent fractions

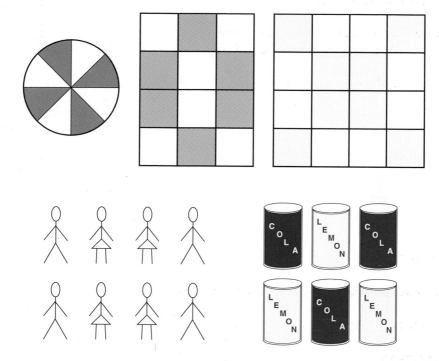

What do all these pictures have in common?

Exercise 14:6

1 Jane folds a piece of paper into four quarters.
  She colours half of the paper red.
  a How many quarters are red?

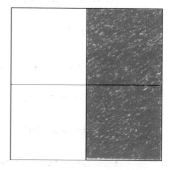

  b Copy this and fill it in.

  $$\frac{1}{2} = \frac{\ldots}{4}$$

2 a How many squares are blue?
   b What fraction of the rectangle is blue?
   c You can write this fraction in more than one way.
     Copy these equal fractions.

$$\frac{3}{6} = \frac{1}{2}$$

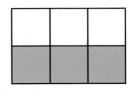

3 What fraction of this rectangle is red?
   Write your answer in two different ways.

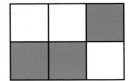

4 Which of these diagrams show a half?

a 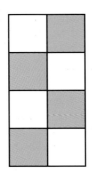   **b** 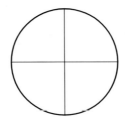   **c**

d 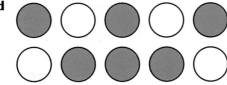   **f**

e 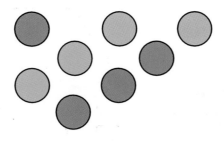   **g**

W 5 Ask your teacher for a copy of worksheet 14 : 1.

6 **a** Look at the shaded squares.
Copy and fill in:

$$\frac{\cdots}{6} = \frac{\cdots}{3}$$

b Look at the unshaded squares.
Copy and fill in:

$$\frac{\cdots}{6} = \frac{\cdots}{3}$$

In questions **7** to **9** write down fractions in the same way as question **6**.

7 **a** Shaded $\dfrac{\cdots}{8} = \dfrac{\cdots}{4}$

b Unshaded $\dfrac{\cdots}{8} = \dfrac{\cdots}{4}$

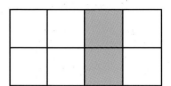

8 **a** Shaded $\dfrac{\cdots}{15} = \dfrac{\cdots}{5}$

b Unshaded $\dfrac{\cdots}{15} = \dfrac{\cdots}{5}$

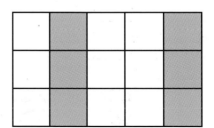

9 **a** Shaded $\dfrac{\cdots}{9} = \dfrac{\cdots}{3}$

b Unshaded $\dfrac{\cdots}{9} = \dfrac{\cdots}{3}$

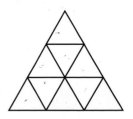

● **10** Copy and fill in the missing numbers.

   a $\dfrac{4}{8} = \dfrac{\cdots}{2}$ **c** $\dfrac{5}{10} = \dfrac{\cdots}{2}$ **e** $\dfrac{10}{20} = \dfrac{\cdots}{2}$

   b $\dfrac{3}{9} = \dfrac{\cdots}{3}$ **d** $\dfrac{6}{9} = \dfrac{\cdots}{3}$ **f** $\dfrac{4}{6} = \dfrac{\cdots}{3}$

3 Using a calculator

Liam and Noel want to change $\frac{3}{8}$ to a decimal.

Liam is having difficulty.
Noel has found the answer on his calculator.

Liam wants to change $\frac{3}{8}$ to a decimal.

Keys to press: 3 ÷ 8 =   0.125

Answer: 0.125

Exercise 14:7

1 Copy this table.
Fill it in.

Fraction	Keys to press	Decimal
$\frac{5}{8}$	5 ÷ 8 =	
$\frac{1}{5}$		
$\frac{3}{4}$		
$\frac{7}{10}$		

2 **a** Change these fractions to decimals: $\frac{2}{5}$ $\frac{1}{4}$
   b Which decimal is larger?
   c Which fraction is larger?

3 For each of these pairs of fractions
   a change the fractions to decimals,
   b use the decimals to help you write down the larger fraction.

   (1) $\frac{4}{5}$ $\frac{7}{8}$ (2) $\frac{14}{25}$ $\frac{6}{10}$ ● (3) $\frac{5}{8}$ $\frac{4}{7}$

4 **a** Change these fractions to decimals: $\frac{5}{8}$ $\frac{10}{16}$
   b Write what you notice.

5 Sort these fractions into pairs of equal fractions.
   Use decimals to help you.

   $\frac{7}{14}$ $\frac{3}{8}$ $\frac{9}{12}$ $\frac{4}{5}$ $\frac{9}{24}$ $\frac{1}{2}$ $\frac{12}{15}$ $\frac{3}{4}$

6 List the fractions that are equal to $\frac{1}{2}$.

   $\frac{7}{14}$ $\frac{6}{12}$ $\frac{5}{10}$ $\frac{3}{8}$ $\frac{10}{18}$ $\frac{50}{100}$

● **7** Sharon has some spanners.
   The sizes of the spanners are
   in fractions of an inch.

   $\frac{3}{8}$ $\frac{1}{2}$ $\frac{3}{4}$ $\frac{5}{8}$ $\frac{7}{16}$

   Sort the spanners into the order
   of their sizes.
   Start with the smallest.

You know that $\frac{4}{8} = \frac{1}{2}$. A calculator with the key $\boxed{a\frac{b}{c}}$ can do this for you.

Example $\frac{4}{8}$ is equal to a simpler fraction. Find this fraction.

Keys to press: $\boxed{4}$ $\boxed{a\frac{b}{c}}$ $\boxed{8}$ $\boxed{=}$ ⌐⌐⌐   Answer $\frac{1}{2}$

Simplify $\frac{4}{8}$ has been **simplified** to $\frac{1}{2}$

Exercise 14:8

Use $\boxed{a\frac{b}{c}}$ to simplify these.

1 $\frac{12}{24}$ **3** $\frac{4}{16}$ **5** $\frac{14}{21}$ **7** $\frac{15}{20}$

2 $\frac{6}{18}$ **4** $\frac{12}{15}$ **6** $\frac{25}{35}$ **8** $\frac{12}{16}$

Example   Simplify $\frac{10}{6}$.

Keys to press: $\boxed{1}$ $\boxed{0}$ $\boxed{a\frac{b}{c}}$ $\boxed{6}$ $\boxed{=}$

The calculator display shows: ⌐⌐⌐⌐   This means $1\frac{2}{3}$.

Answer $\frac{10}{6} = 1\frac{2}{3}$

Exercise 14:9

Write these calculator displays as whole numbers and fractions.

1 ⌐⌐⌐⌐   **2** ⌐⌐⌐⌐   **3** ⌐⌐⌐⌐   **4** ⌐⌐⌐⌐

Use $\boxed{a\frac{b}{c}}$ to simplify these.

5 $\frac{8}{3}$ **7** $\frac{22}{8}$ **9** $\frac{16}{3}$ **11** $\frac{19}{6}$

6 $\frac{14}{4}$ **8** $\frac{22}{5}$ **10** $\frac{14}{7}$ **12** $\frac{24}{8}$

Example   Find $\frac{3}{4}$ of 44.

Use ✗ for 'of' because 'of' means multiply.

If your calculator has a $a\frac{b}{c}$ key, press these keys:

3 $a\frac{b}{c}$ 4 ✗ 4 4 = ꓱꓱ

If your calculator does not have a $a\frac{b}{c}$ key, press these keys:

3 ÷ 4 ✗ 4 4 = ꓱꓱ

Answer 33

Exercise 14:10

Use your calculator to work these out.

1 $\frac{2}{3}$ of 24   **3** $\frac{5}{6}$ of 18   **5** $\frac{9}{10}$ of 30

2 $\frac{3}{5}$ of 30   **4** $\frac{3}{4}$ of 36   **6** $\frac{3}{8}$ of 40

7 There are 28 pupils in 9M.
Three quarters of the class eat lunch at school.
a Write the fraction in figures.
b How many pupils eat lunch at school?

8 Parvinda has asked some Year 9 pupils how many pieces of fruit they eat.
Two thirds of pupils eat at least one piece of fruit every day.
a Write the fraction in figures.
b There are 36 pupils in Parvinda's survey.
How many eat at least one piece of fruit every day?

9 Four fifths of the girls in Year 9 at Stanthorne High like to play tennis.
If there are 90 girls, how many like to play tennis?

14

Exercise 14:11

Mountain bike £198
Helmet £27
Cycle bottle £1.95
Cycling route book £3.99
Cycling jersey £15.99
Cycling shorts £18
Cycling gloves £4.50

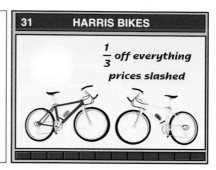

31   HARRIS BIKES

$\frac{1}{3}$ off everything

prices slashed

Example Ben wants the cycling helmet.

 a How much does he save on the helmet?
 b What does he pay for the helmet?

 a $\frac{1}{3}$ of 27 = 27 ÷ 3 = 9 Ben saves £9

 b £27 − £9 = £18 Ben pays £18 for the helmet.

In questions **1** to **6** the prices are next to the picture.
For each of these work out
 a how much you save,
 b the new price.

1 Cycling shorts   **3** Cycling gloves   **5** Cycle bottle

2 Mountain bike   **4** Cycling route book   **6** Cycling jersey

● **7** Danny is going to the sports centre.
 All prices are reduced by $\frac{1}{4}$.
 Danny is going to the gym first and then he is going swimming.
 Swimming usually costs £1.20 and the gym usually costs 80 p.
 How much will Danny pay in total at the reduced price?

1 **a** What do we divide by to find $\frac{1}{5}$ of a number?

Find: **b** $\frac{1}{5}$ of 10 **c** $\frac{1}{5}$ of 15 **d** $\frac{1}{5}$ of 50

2 **a** What do we divide by to find $\frac{1}{10}$ of a number?

Find: **b** $\frac{1}{10}$ of 20 **c** $\frac{2}{10}$ of 20 **d** $\frac{3}{10}$ of 20

3 **a** Share seven apples equally between three children.

b Share three cakes equally between four children.

4 Which of these diagrams show a third?

a 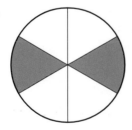 **c**   **e**

b  **d**  **f**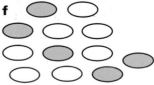

5 Write each of these pictures
   a as a 'top heavy fraction',
   b as a whole number and a fraction.

(1) 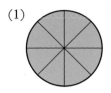   (2)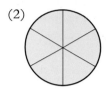

6 **a** Change these fractions to decimals: $\frac{3}{8}$ $\frac{5}{16}$
   b Which decimal is larger?
   c Which fraction is larger?

7 Which of these fractions are equal to $\frac{1}{4}$?
   $\frac{2}{8}$ $\frac{3}{16}$ $\frac{6}{24}$ $\frac{3}{12}$ $\frac{30}{40}$ $\frac{4}{16}$ $\frac{21}{80}$ $\frac{25}{100}$

8 Work these out.
   a $\frac{2}{3}$ of 18   **c** $\frac{1}{8}$ of 32   **e** $\frac{7}{10}$ of 30
   b $\frac{3}{4}$ of 24   **d** $\frac{4}{5}$ of 35   **f** $\frac{5}{6}$ of 12

9 Last season the school football team played 21 matches.
   They won two thirds of the matches.
   How many did they win?

10 One quarter of the drinks sold by this
   drinks machine contain sugar.
   The machine sold 520 drinks in one week.
   How many of these contained sugar?

11 Stanthorne High pupils have had their GCSE results.
   Nine tenths of the pupils passed.
   If 140 pupils sat GCSE, how many passed?

12 a This flag has four stripes.
What fraction of the flag is the red stripe?

b Draw a rectangle on **plain** paper.
Divide your rectangle into quarters like the flag by *estimating*.
Colour the stripes.

c Draw another rectangle on **plain** paper.
Divide this rectangle into thirds.
Colour the stripes.

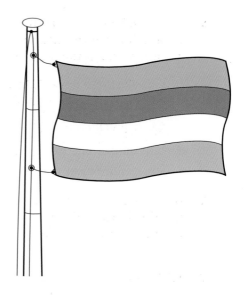

13 The diagram shows a swimming pool from above.
The lines mark the swimming 'lanes'.
The lanes have numbers from **1** to **5**.

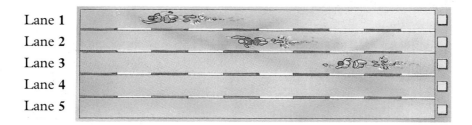

Lane 1
Lane 2
Lane 3
Lane 4
Lane 5

a Which number lane has a swimmer who is half way to the finish?

b Which number lane has a swimmer who is a quarter of the way to the finish?

c Which number lane has a swimmer who is three quarters of the way to the finish?

d Sketch the two empty lanes.
In lane **4** draw a swimmer one third of the way to the finish.
In lane **5** draw a swimmer two thirds of the way to the finish.

1 **a** Use a calculator to change $\frac{1}{3}$ to a decimal.
Write down all the figures on the calculator display.

b Repeat part **a** for $\frac{2}{3}$.

c Change these fractions to decimals using a calculator.

$$\frac{1}{9} \quad \frac{2}{9} \quad \frac{3}{9} \quad \frac{4}{9}$$

Write down your answers.

d Use the pattern to write down the decimals for $\frac{5}{9}$ $\frac{6}{9}$ $\frac{7}{9}$ and $\frac{8}{9}$.

e What number of ninths has the same pattern as $\frac{1}{3}$?

f What number of ninths has the same pattern as $\frac{2}{3}$?

g Investigate the decimal patterns for $\frac{1}{11}$ $\frac{2}{11}$ etc.

2 Between each pair of fractions insert one of these symbols:
< (less than), = or > (greater than).

a $\frac{1}{2}$ $\frac{5}{8}$ **b** $\frac{14}{21}$ $\frac{2}{3}$ **c** $\frac{6}{8}$ $\frac{15}{20}$ **d** $\frac{5}{12}$ $\frac{6}{24}$

3 Copy this grid.

a Colour half the squares red.
Write down the fraction that are not red.

b Colour a quarter of the squares blue.

c Colour one square green.

d Write down the fraction of the rectangle that is green.

e Write down the fraction of the squares that have not been coloured.

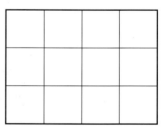

4 1260 people attended a pop concert.
Two thirds of the people were under 20.

a How many people were under 20?

b What fraction of the people were 20 or older?

- *Example*   Phil needs to find $\frac{1}{5}$ of 10.

  To find $\frac{1}{5}$ divide by 5: $10 \div 5 = 2$ $\frac{1}{5}$ of 10 is 2

- Liam wants to change $\frac{3}{8}$ to a decimal.

  Keys to press: **3** **÷** **8** **=**   Answer: 0.125

- You know that $\frac{4}{8} = \frac{1}{2}$. A calculator with the key **$a\frac{b}{c}$** can do this for you.

  Example $\frac{4}{8}$ is equal to a simpler fraction. Find this fraction.

  Keys to press: **4** **$a\frac{b}{c}$** **8** **=** $1 \rfloor 2$ Answer: $\frac{1}{2}$

  Simplify: $\frac{4}{8}$ has been **simplified** to $\frac{1}{2}$.

- You can use a calculator to simplify fractions.

  Example   Simplify $\frac{10}{6}$.

  Keys to press: **1** **0** **$a\frac{b}{c}$** **6** **=**

  The calculator display shows: $1 \rfloor 2 \rfloor 3$ This means $1\frac{2}{3}$.

  Answer $\frac{10}{6} = 1\frac{2}{3}$

- *Example*   Find $\frac{3}{4}$ of 44.

  If your calculator has a **$a\frac{b}{c}$** key, press these keys:

  3 **$a\frac{b}{c}$** **4** **×** **4** **4** **=** 33

  If your calculator does not have a **$a\frac{b}{c}$** key, press these keys:

  3 **÷** **4** **×** **4** **4** **=** 33

  Answer 33

- *Example*   A cycling helmet costs £27.
  There is $\frac{1}{3}$ off the price in a sale.
  Ben wants to buy the helmet.
  a How much does he save on the helmet?
  b What does he pay for the helmet?

  a $\frac{1}{3}$ of $27 = 27 \div 3 = 9$ Ben saves £9

  b £27 − £9 = £18   Ben pays £18 for the helmet.

1 Find: **a** $\frac{1}{2}$ of 20   **b** $\frac{1}{4}$ of 20   **c** $\frac{3}{4}$ of 20

2 Share four pizzas between three children.

3 **a** Which of these fractions does not simplify to one half?

$\frac{1}{2}$ $\frac{2}{4}$ $\frac{2}{8}$ $\frac{3}{6}$ $\frac{6}{12}$ $\frac{8}{16}$

b Which of these fractions does not simplify to one third?

$\frac{1}{3}$ $\frac{2}{6}$ $\frac{3}{9}$ $\frac{4}{12}$ $\frac{10}{30}$ $\frac{12}{33}$

4 Change $\frac{8}{25}$ to a decimal.

5 Which fraction is larger $\frac{2}{5}$ or $\frac{3}{10}$?

6 Work out $\frac{3}{4}$ of 28.

7 Two fifths of the chocolates in this box are covered in white chocolate.
There are 30 chocolates in the box.
  a What fraction of the chocolates are white? Write this fraction in figures.
  b How many chocolates are white?

8 A pair of shoes costs £20.40
The price is reduced by one quarter.
  a What do you save if you buy the shoes?
  b What is the new price of the shoes?

15 The best chapter, probably

QUESTIONS

EXTENSION

SUMMARY

TEST YOURSELF

A hurricane is an intense, devastating tropical storm caused by a low-pressure weather system. Hurricanes occur in tropical regions usually between July and October.

Weather forecasters try to warn people of the approach of a hurricane. The forecasters use probabilities to give an idea of the likelihood of a hurricane striking a particular area.

1 Beads and spinners

There are 10 beads in a bag.
Some of the beads are red and the rest are black.

David and Pavneet have to find out how many beads of each colour are in the bag.
They must not look in the bag.

David and Pavneet take one bead out of the bag.

They note its colour and replace it.
They do this lots of times.

Exercise 15:1

An experiment to find the number of beads of each colour in a bag.

You need a bag containing 10 coloured beads.

You are going to find out the colour of the beads.
You are also going to find out how many there are of each colour.
You are going to do this *without looking*.

1 **a** Copy this tally-table.
Use it to record your results.

Colour	Tally	Total

b Take one bead out of the bag.
Record its colour in the tally-table.
Put the bead back in the bag.
Do this 10 times.
Put the bead back each time.
You should have 10 tally marks in your tally-table.

c Complete the 'Total' column of the tally-table.

2 **a** Make another copy of the tally-table like the one for question **1**.
b Take a bead out of the bag and record its colour.
Put the bead back in the bag.
Repeat the experiment 10 times as you did before.
c Complete the 'Total' column of your tally-table.

3 Make another copy of the tally-table.
Repeat the experiment another 10 times.

4 Look at the 'Total' column of your three tally-tables.
   a Are the totals for each colour the same each time?
   b Do you think you know the number of beads of each colour in the bag?
   If you answer:
   'no', make a new tally-table and repeat the experiment another 10 times,
   'yes', write down what you think is in the bag.

5 When you think you know the number of beads of each colour in the
bag, look in the bag.
Did you get the answer right?
If you got the answer wrong, write down a reason.

Exercise 15:2

1 Sally puts these beads in a bag:

She then takes out a bead without looking.
   a Which colour bead is Sally most likely to get?
   b Which colour bead is Sally least likely to get?

2 Lil puts these counters in a bag:

She takes out a counter without looking.
   a Which colour counter is Lil more likely to get?
   b Lil wants to make it **equally likely** that she will pick a red counter
   or a blue counter.
   What extra counters does Lil need to add to her bag?

3 There are 10 beads in a bag.
Andrew takes out a bead without looking.
He records the colour and puts the
bead back in the bag.
Andrew does this 10 times.

Here is Andrew's tally-table:

Colour	Tally	Total
red	ＷＩ	5
blue	ＩＩＩ	3
yellow	ＩＩ	2

a Andrew says,
'There must be 5 red beads in my bag because there are
5 reds in my table.'
Explain why Andrew is wrong.
b What is the smallest number of yellow beads there could be
in the bag?
c Andrew takes out another bead from his bag.
What colour is the bead most likely to be?

4 Harneet has 20 cubes in a bag.
She takes out a cube without looking.
Harneet writes down the colour and puts the cube back.
Harneet does this 20 times.

Harneet records her
results in this chart:

Green	4
Yellow	5
White	8
Blue	3

a Harneet says,
'There are 4 greens in my table so there must be 4 green cubes
in the bag.'
Explain why Harneet is wrong.
b What is the smallest number of blue cubes that there could be in
the bag?
c Harneet says,
'There cannot be any red cubes in my bag because there are no
reds in my table.'
Explain why Harneet is wrong.

Danny knows that his bag contains 10 counters, 7 red and 3 yellow.

Danny picks a counter from the bag without looking.

He writes down the colour and puts it back in the bag.

Danny does this 10 times.

He *expects* to get 7 reds and 3 yellows.

Exercise 15:3

1 Danny picks a counter from the same bag 20 times.
   a How many yellows does he expect to get?
   b How many reds does he expect to get?

2 Danny picks a counter from the same bag 40 times.
   a How many yellows does he expect to get?
   b How many reds does he expect to get?

3 Danny picks a counter from the same bag 50 times.
   a How many yellows does he expect to get?
   b How many reds does he expect to get?

4 This dice is rolled a number of times.
   Write down the number of sixes you expect in:
   a 6 throws   **b** 12 throws   **c** 60 throws

5 A spinner has 4 equal sections.
   How many greens do you expect if you spin it:
   a 4 times   **c** 40 times
   b 8 times   ● **d** 100 times?

2 How much chance?

Paul and Liz are playing 'Flounders'.

Paul needs to win.

There are **six** numbers on the dice.

Paul has a $\frac{1}{6}$ chance of getting the

 that he wants.

The probability is $\frac{1}{6}$.

Exercise 15:4

1 Liz throws the dice.
Write down the probability that she gets:

a   b   c

2 Manuel has five tins of soup, but the labels have come off.

He knows that he bought three tins of tomato and two tins of chicken.
Manuel opens a tin.
Write down the probability that it contains:
a tomato soup
b chicken soup
c mushroom soup

3 This box contains 8 counters.
The counters are either red or blue.
The probability of picking a red
counter without looking is $\frac{3}{8}$.

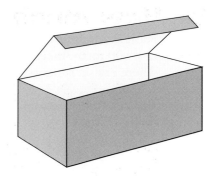

   a How many red counters are in the box?
   b How many blue counters are in the box?

4 Mrs Patel has 4 black beads and
1 orange bead in a bag.

She asks 9P the probability of
picking an orange bead without
looking.

Tracy says that it is $\frac{1}{4}$ because there is 1 orange bead and 4 black beads.
Michael says that it is $\frac{1}{5}$ because there are 5 beads and 1 is black.

   a Which answer is correct?
      Explain why the other answer is wrong.
   b What is the probability of picking a black bead without looking?

5 What is the probability of getting red on each of these spinners?
  a   **b**   **c**

6 What is the probability of getting a 4 on each of these spinners?
  a   **b**   **c**

 7 Ask your teacher for a copy of worksheet 15:2.

Katie is running
in a 100 metre race.
There are six runners.

The runners are **not all equally
likely** to win.

The probability that Katie
wins is **not** $\frac{1}{6}$.

The probability depends on
how fast Katie can run.

Exercise 15:5

Decide whether each of these is true or false.

1 The school canteen sells milk or cola.
There are 2 drinks so the probability of a person having cola is $\frac{1}{2}$.

2 A coin has 2 sides, heads or tails.
The probability of getting tails if you toss the coin is $\frac{1}{2}$.

3 In the summer, Year 9 pupils can choose to do athletics, cricket or tennis.
There are 3 sports so the probability that a pupil chooses tennis is $\frac{1}{3}$.

4 This spinner has 3 sections.
The probability of getting blue is $\frac{1}{3}$.

5 A shop sells balls in 3 colours, red, blue and yellow.
The probability of a person choosing a red ball is $\frac{1}{3}$.

6 A cafe sells two hot snacks, hot dogs and beefburgers.
David wants a hot snack.
The probability that he chooses a beefburger is $\frac{1}{2}$.

3 Probability methods and scales

Karen and Rob want an estimate of the probability that it will rain on the day of the school fair in June.

They need to look at *data* for the weather in June in past years.

There are three methods of finding probabilities.

Method 1 Use equally likely outcomes

e.g. the probability of getting a 4 on a dice is $\frac{1}{6}$.

Method 2 Use a survey or do an experiment
e.g. to find the probability that a car passing the school is red, do a survey of the colours of cars passing the school.

Method 3 Look back at data
e.g. to find the probability that it will rain on a day in June, look at the number of days it rained in June in past years.

Exercise 15:6

Look at these situations.
Say whether you would use *method 1*, *method 2* or *method 3* to estimate each of these probabilities.

1 The probability that you win a raffle if you buy one ticket and there are 100 tickets sold.

2 The probability that a dropped drawing pin will land with its pin up.

3 The probability that a pupil chosen at random from your school had cereal for breakfast.

4 The probability that it will snow in London on 25th December this year.

5 The probability that you will get red on this spinner:

6 The probability that from a choice of athletics, cricket or tennis, a pupil will choose tennis.

· ·

We can show probabilities on a **probability scale**:

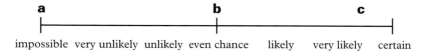

impossible very unlikely unlikely even chance likely very likely certain

a The grass on the school field will turn purple tomorrow.
b If I toss a coin it will land heads up.
c It will be a sunny day in the Sahara desert.

We often draw probability scales with a number scale:

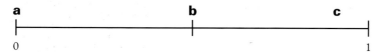

An event that is **impossible** has a **probability of 0**.
An event that is **certain** has a **probability of 1**.

Exercise 15:7

1 Copy this probability scale.

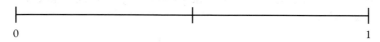

Mark these probabilities on your scale.
a It will snow in Manchester in January.
b It will snow in Manchester in April.
c It will snow in Manchester in August.

2 a A Year 9 pupil is chosen at random.
How likely is it that the pupil has brought each of these to school?
Write the objects in order starting with the least likely.

b Copy this probability scale.

Mark the probabilities that the pupil has brought these on your scale.
The first one has been done for you.

• 3 This drinks machine is broken.
You cannot choose the drink you get!

Jane likes all the drinks.
Paul only likes the chicken soup.

Jane and Paul buy one drink each.

Copy this probability scale.

Mark these probabilities on your scale:
a Jane will get a drink that she likes.
b Paul will get a drink that he likes.
c Ann buys a drink. The arrow shows the probability that Ann gets a drink that she likes.

How many drinks does Ann like?

1   Harriet puts these beads into a bag.
She picks out a bead without looking.
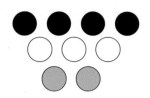
a   Which colour is Harriet most likely to choose?
b   Which colour is Harriet least likely to choose?
c   Harriet wants all the colours to have an
equal chance.
What beads does Harriet need to add to
her bag?

2   Carol puts 12 counters in a bag.
She chooses a counter at random and then replaces it.
Carol records the colour of the counter she chooses.
She does this 12 times.
Here is Carol's tally-table:

Colour	Tally	Total
green	llll l	6
yellow	llll	4
red	ll	2

a   Carol says,
'There must be 6 green counters in my bag because there are
6 greens in my table'.
Explain why Carol is wrong.
b   What is the smallest number of red counters that can be in the bag?
c   Carol says,
'There cannot be any blue counters in my bag because there are
no blues in my table.'
Explain why Carol is wrong.

3   Peter tosses a 2 p coin lots of times.
Write down the number of heads he
expects when he tosses the coin:

a   twice   **c**   60 times
b   10 times   **d**   100 times

4   A fair dice is rolled.
Write down the probability of getting each of these:
a   **b**   **c**

5 Tara puts these beads into a bag.
She asks her friends the probability of
choosing the red bead without looking.

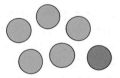

Ailsa says it is $\frac{1}{6}$ because there are 6 beads and 1 is red.

Zoe says it is $\frac{1}{5}$ because there are 5 blue beads and 1 red bead.

a Whose answer is correct?
Explain why the other answer is wrong.

b What is the probability of choosing a blue bead?

6 Write down the method you would use to estimate the probability of
each of these.

Choose from: equally likely outcomes
                    a survey or an experiment
                    looking back at data

a The probability of spinning red on this spinner.
b The probability of tossing a 6 on a biased dice.
c The probability of a sunny day in July.

7 9M are doing a survey of the traffic on the road outside the school.
How likely is it that each of these will pass the school?

a Write them in order starting with the least likely.

b Copy this probability scale.

0   1

c Show the probability of each vehicle on your scale.

1   Mandy has 10 beads of three different colours in a bag.
    She wants to find out how many beads there are of each colour
    without looking.
    Mandy takes out a bead and writes down the colour.
    Mandy then puts the bead back in the bag.
    How many times does Mandy need to repeat this to predict the
    number of each colour?
    Explain your answer.

2   Sally spins this spinner 12 times:
    a How many reds does Sally
         expect to get in 12 spins?
    b How many blues does Sally
         expect to get in 12 spins?

    Here are Sally's results:

Colour	Tally	Total
red	|	1
blue	ЖІ	5
green	|||	3
yellow	|||	3

    c Sally thinks that the spinner is not fair because she has
         1 red and 5 blues.
         How can Sally find out if the spinner is fair?

3   **a** Fay puts these counters in a bag:
         Fay goes to pick a counter
         without looking.
         Fay says that the probability of
         getting a red is $\frac{1}{3}$ because red is
         1 colour and there are 3 colours
         in the bag.
         Explain why Fay is wrong.

    b The school tuck shop sells these crisps: salt and vinegar,
         smoky bacon, cheese and onion.
         Chris says that as there are 3 flavours, the probability of a person
         buying cheese and onion is $\frac{1}{3}$.
         Explain why Chris is wrong.

- Danny knows his bag contains 10 counters, 7 red and 3 yellow.
  Danny picks a counter from the bag without looking.
  He writes down the colour and puts it back in the bag.
  Danny does this 10 times.
  He *expects* to get 7 reds and 3 yellows.

- There are **six** numbers on a dice.

  Paul has a $\frac{1}{6}$ chance of getting the

  The probability is $\frac{1}{6}$.

- Katie is running
  in a 100 metre race.
  There are six runners.

  The runners are **not all equally likely** to win.

  The probability that Katie wins is **not** $\frac{1}{6}$.

  The probability depends on how fast Katie can run.

- There are three methods of finding probabilities.
  Method 1   Use equally likely outcomes.
  Method 2   Use a survey or do an experiment.
  Method 3   Look back at data.

- We can show probabilities on a **probability scale**:

  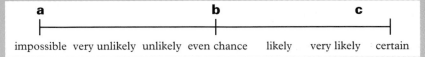

  a The grass on the school field will turn purple tomorrow.
  b If I toss a coin it will land heads up.
  c It will be a sunny day in the Sahara desert.

  We often draw probability scales with a number scale:

  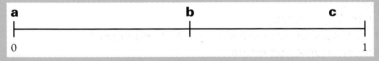

  An event that is **impossible** has a **probability of 0**.
  An event that is **certain** has a **probability of 1**.

1   Lucy puts these cubes in a box:
     She takes out a cube without looking.

   a   Which colour is Lucy more likely to get?
   b   Which colour is Lucy less likely to get?
   c   Write down the probability of getting a red.
   d   Write down the probability of getting a blue.
   e   Write down the probability of getting a yellow.

2   Rashid spins this spinner:
     He writes down the colour.
     How many reds does he expect
     to get if he spins the spinner:
     a   3 times
     b   9 times
     c   30 times?

3   Decide whether each of these is true or false:
     a   Year 9 can choose to study French or German.
         There are 2 choices so the probability that a pupil chooses French
         is $\frac{1}{2}$.
     b   An ordinary dice is thrown. There are 6 numbers so the probability
         of getting a 5 is $\frac{1}{6}$.

4   What method would you use to find the probability that one of Year 9
     chosen at random owns a bicycle?
     Choose from: equally likely outcomes, survey, looking at data.

5   Copy this probability scale.

0   1

   Mark these probabilities on your scale:
   a   Getting a head when a 2 p coin is tossed.
   b   Somebody in your maths class has forgotten their exercise book.
   c   You would get into trouble if you were rude to your headteacher.
   d   The water in the Atlantic Ocean will dry up tomorrow.

16 Percentages

The percent symbol (%) seems to have originated in business.

In the sixteenth century a symbol like a capital Z with a circle at either end was written to represent one one-hundredth.

1 Percentages

71% of the surface of the Earth is covered by water.

This pie-chart has 100 parts.
Each part is 1% of the pie-chart.

7% is green.
$100\% - 7\% = 93\%$

93% is not green.

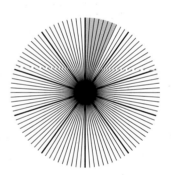

Exercise 16:1

For each diagram write down:
a   the percentage that is coloured,
b   the percentage that is not coloured.

1

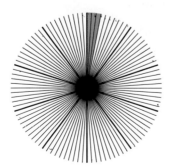

2

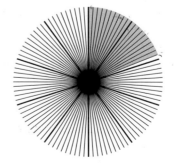

3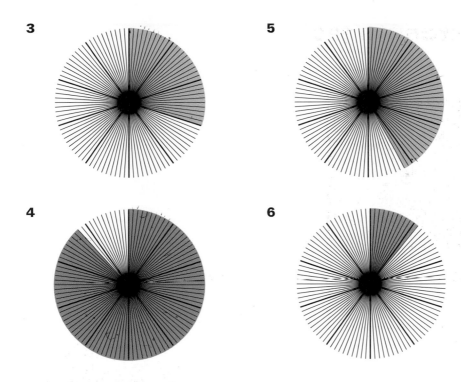

5

4

6

7 The human body is about 70% water.
What percentage is not water?

8 Cara has done 82% of her homework.
What percentage does she still have to do?

9 Mark has coloured in
35% of his picture.
What percentage does
he still have to colour?

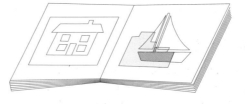

10 12% of Year 9 at Stanthorne High go home to lunch.
What percentage stay at school to lunch?

11 The percentage of homes that have a telephone is 87%.
What percentage of homes do not have a telephone?

12 Sally is peeling potatoes.
She cuts off 8% as peel.
What percentage of the potatoes is left?

1% of this pie-chart is coloured red.

1% = $\frac{1}{100}$ is coloured red.

17% is coloured blue.

The fraction coloured blue is $\frac{17}{100}$.

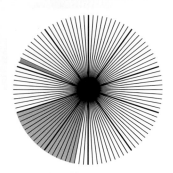

Exercise 16:2

1 Copy these and fill in the fractions.

a $23\% = \dfrac{...}{100}$ **d** $91\% = ...$ **g** $11\% = ...$

b $43\% = ...$ **e** $7\% = ...$ **h** $9\% = ...$

c $79\% = ...$ **f** $67\% = ...$ **i** $41\% = ...$

2 Write these fractions as percentages.

a $\dfrac{38}{100}$ **c** $\dfrac{5}{100}$ **e** $\dfrac{8}{100}$ **g** $\dfrac{16}{100}$

b $\dfrac{77}{100}$ **d** $\dfrac{65}{100}$ **f** $\dfrac{99}{100}$ **h** $\dfrac{44}{100}$

Sometimes you can simplify the fractions.

10% is blue.

$10\% = \dfrac{10}{100}$

$\dfrac{10}{100}$ in its simplest

form is $\dfrac{1}{10}$

so $10\% = \dfrac{1}{10}$.

$\frac{10}{100}$ is blue

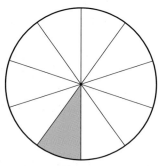

$\frac{1}{10}$ is blue

Exercise 16:3

1　**a**　How many parts are green?
　　b　What percentage is green?
　　c　Write this percentage as a fraction of 100.
　　d　Write your fraction in its simplest form.

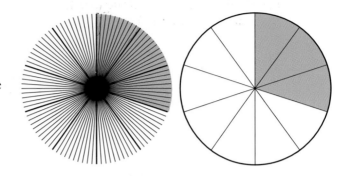

2　Copy these and fill them in.

　　a　$70\% = \dfrac{\ldots}{100} = \dfrac{\ldots}{10}$　　　　　　**b**　$90\% = \dfrac{\ldots}{100} = \dfrac{\ldots}{10}$

3　**a**　How many parts are red?
　　b　What percentage is red?
　　c　Write this percentage as a fraction of 100.
　　d　Write your fraction in its simplest form.

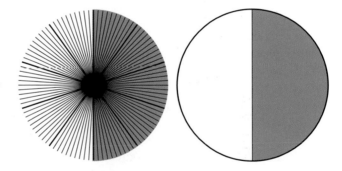

4　**a**　How many parts are blue?
　　b　What percentage is blue?
　　c　Write this percentage as a fraction of 100.
　　d　Write your fraction in its simplest form.

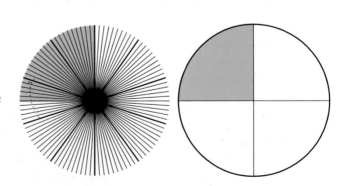

5　Copy this and fill it in.

　　$75\% = \dfrac{\ldots}{100} = \dfrac{\ldots}{4}$

6 **a** What percentage is
purple?
  b Write this percentage
as a fraction of 100.
  c Write your fraction
in its simplest form.

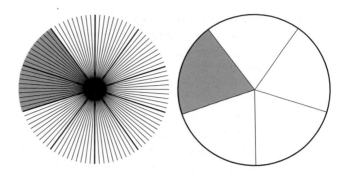

7 Copy these and fill them in.

  a $40\% = \dfrac{\cdots}{100} = \dfrac{\cdots}{5}$

  b $60\% = \dfrac{\cdots}{100} = \dfrac{\cdots}{5}$

  c $80\% = \dfrac{\cdots}{100} = \dfrac{\cdots}{5}$

8 Kevin buys 3 red pens and 7 blue pens.
  a What fraction of the pens are red?
  b What percentage of the pens are red?
  c What fraction of the pens are blue?
  d What percentage of the pens are blue?

9 Katie has made patterns with squares.
  a What fraction of the squares are green?
  b What percentage of the squares are green?
  c What fraction of the squares are yellow?
  d What percentage of the squares are yellow?

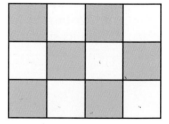

10 Jennie has made this shape with cubes.
  a What fraction of the cubes are red?
  b What percentage of the cubes are red?
  c What fraction of the cubes are yellow?
  d What percentage of the cubes are yellow?

11 **a** What fraction of the squares are red?
  b What percentage of the squares are red?
  c What fraction of the squares are yellow?
  d What percentage of the squares are yellow?
  e What fraction of the squares are green?
  f What percentage of the squares are green?

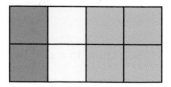

12 **a** What fraction of these drinks are orange?
b What percentage of the drinks are orange?
c What fraction of the drinks are apple?
d What percentage of the drinks are apple?
e What fraction of the drinks are blackcurrant?
f What percentage of the drinks are blackcurrant?

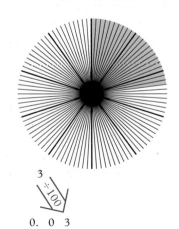

● **13** Write these fractions as percentages.

a $\dfrac{1}{2}$ **c** $\dfrac{3}{4}$ **e** $\dfrac{3}{10}$ **g** $\dfrac{1}{5}$ **i** $\dfrac{3}{5}$

b $\dfrac{1}{4}$ **d** $\dfrac{1}{10}$ **f** $\dfrac{7}{10}$ **h** $\dfrac{2}{5}$ **j** $\dfrac{4}{5}$

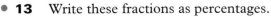

23% of this pie-chart is blue.

The fraction that is blue is $\dfrac{23}{100}$.

You can change $\dfrac{23}{100}$ to a decimal.

$\dfrac{23}{100} = 23 \div 100 = 0.23$

Example   Change $\dfrac{3}{100}$ to a decimal.

$3 \div 100 = 0.03$

3
÷100
0. 0 3

Exercise 16:4

Write each of these percentages as:
a a fraction of 100,
b a decimal.

1 17%   **4** 8%   **7** 9%   **10** 6%

2 85%   **5** 1%   **8** 2%   **11** 12%

3 42%   **6** 5%   **9** 4%   **12** 73%

2   Percentage of an amount

· ·

Lisa wants to buy the tent which costs £120. She wonders how much
she can save.

25% is the same as $\frac{1}{4}$.

You divide by 4 to find 25%.

Example

Lisa is buying a tent.
Find how much Lisa saves.

$$\frac{1}{4} \text{ of } £120 = £120 \div 4$$

$$= £30$$

Lisa saves £30.

Exercise 16:5

1   Use the prices in the picture for this question.
How much will Lisa save on:

    a   a water carrier   **e**   two airbeds
    b   a gas cooker   **f**   two sleeping bags
    c   a lantern   **g**   a camping mat
    d   a folding table and stool set   **h**   a cool box?

A sports shop reduces all its prices by 10%.

> **SUPER SPORTS SALE**
>
> Trainers £60   Tennis racket £45
> Fishing rod £20   Cricket bat £26
> Tackle box £12   Tennis balls £7
> Football £10   Cycle helmet £24.80
> Football goal £24   Cycle computer £16.80
> Football pump £6.50   Squash balls £5
> Dartboard £12.70   Cycle lights £10.40
> Darts £8

10% is the same as $\dfrac{1}{10}$.
You divide by 10 to find 10%.

Example

Jane is buying a dart board and some darts.
Find how much Jane saves on:

a a dart board   **b** some darts

a $\dfrac{1}{10}$ of £12.70 = £12.70 ÷ 10

    = £1.27

Jane saves £1.27

```
1 2 . 7 0
   ÷10
1 . 2 7
```

b $\dfrac{1}{10}$ of £8 = £8.00 ÷ 10

    = 80 p

Jane saves 80 p

```
8 . 0 0
  ÷10
. 8 0
```

---

2 Use the prices in the picture for this question.
How much would you save on:

a a football
b a fishing rod
c cycle lights
d a cycle helmet
e a football pump

f tennis balls
g squash balls
h a tackle box
i a football goal
j a tennis racket?

3 A new mountain bike costs £180.
The shop is selling a second-hand bike at 50% off the full price.
How much would Hank save if he bought this bike?

Remember: 50% is the same as $\dfrac{1}{2}$

4 Huw gets £10 per week for his paper round.
The shop keeper is going to increase Huw's money by 10%.
How much extra will Huw get?

5 A packet of sweets contains 200 g.
A special offer pack contains 20% more.
How many extra grams of sweets is this?

Remember: 20% is the same as $\dfrac{1}{5}$

6 A box of tea contains 160 teabags.
A special offer pack contains 10% more.
a How many extra teabags is this?
b How many teabags are there altogether in the special offer pack?

You may need to use a calculator to work out a percentage.

Example   Sian is buying a computer.
It costs £960.
The shop reduces the price by 12%.
How much does Sian save?

Sian needs to work out 12% of 960.

	Step 1	**Step 2**	**Step 3**	**Step 4**
This can be written:	$\dfrac{12}{100}$	×	960	=

Step 1 This changes the percentage to a decimal.
The decimal appears when you press ☒ at step 2.
Step 2 'Of' is the same as multiply so use the ☒ key.
Step 3 You are finding the percentage of this amount.
Step 4 This gives you the answer.

Keys to press:

Step 1   **Step 2**   **Step 3**   **Step 4**

1 2 ÷ 1 0 0   ☒   9 6 0   =

Answer £115.20

Exercise 16:6

1 Find 12% of 300 g.
   a Copy and fill in:

   $$12\% \text{ of } 300 = \frac{12}{\cdots} \times 300$$

   `12` `2ndF` `%` `=` `×` `300`
   to get %

   b Use your calculator to work out the answer.

2 Find 18% of 250 cm.
   a Copy and fill in:

   $$18\% \text{ of } 250 = \frac{\cdots}{100} \times \cdots$$

   b Use your calculator to work out the answer.

3 Find 24% of 350 m.
   a Copy and fill in:

   $$24\% \text{ of } 350 = \frac{\cdots}{\cdots} \times \cdots$$

   b Use your calculator to work out the answer.

4 Find 8% of 600 cm

5 Find 34% of 350 m

6 Find 15% of £450

7 Find 3% of 17 kg

8 Find 40% of 120 g

9 Find 30% of £150

10 Find 5% of 420 cm

11 Find 28% of £725

12 Find 19% of 250 t

13 Find 12% of 24 m

14 There are 40 ink cartridges in a packet.
   This special offer packet contains 15% more.
   a How many extra cartridges is this?
   b How many cartridges are there altogether?

Pen Cartridges
15% extra
Free!
Writes for miles and miles more

Exercise 16:7

1 The volume of a standard size bottle of bath oil is 300 ml.
A special offer bottle contains 8% more.
How much more bath oil is in the new bottle?

2 There are 900 pupils in Lisa's school.

    a 44% of the pupils have blood group A.
       How many of the pupils have blood group A?

    b 4% of the pupils have blood group AB.
       How many of the pupils have blood group AB?

3 Helen's family went out for a meal.
The bill came to £32.50
They gave the waiter a 12% tip.
How much was the tip?

4 The volume of a packet of coffee is 660 cm³.
A larger size packet holds 40% more.

    a How much more coffee is in the larger packet?

    b How much coffee is there in the larger packet altogether?

5 Marmalade is made from 30% oranges, 55% sugar and the rest is
water.

    a What percentage is water?

    b A jar contains 340 g of marmalade.
       Find the number of grams of:
       (1) oranges   (2) sugar   (3) water

6 Michael says that 15% of 60 is the same as 60% of 15.
Is Michael correct?
Show your working.

1 Todd and his friends are racing model motor boats.

Start   Finish

1

2

3

4

   a Which lane has a boat that is 50% of the way to the finish?
   b Which lane has a boat that is 25% of the way to the finish?
   c Which lane has a boat that is 75% of the way to the finish?
   d Sketch the empty lane.
      Draw a boat that is 90% of the way to the finish.

2 Copy these.
   Fill in the missing numbers.

   a $20\% = \dfrac{...}{100} = \dfrac{1}{...}$ **c** $25\% = \dfrac{...}{100} = \dfrac{1}{...}$

   b $30\% = \dfrac{...}{100} = \dfrac{3}{...}$ **d** $60\% = \dfrac{...}{100} = \dfrac{...}{10} = \dfrac{3}{...}$

3 Paula wants to buy a personal stereo.
   She has saved up 65% of the money.
   What percentage does she still have to save?

4 48% of the pupils in Susan's form are boys.
   What percentage are girls?

5 John, Louise and Gita are delivering leaflets.
   John delivers 37%.
   Louise delivers 32%.
   What percentage does Gita deliver?

6 Change these fractions to percentages:

   a $\dfrac{40}{100}$ **b** $\dfrac{19}{100}$ **c** $\dfrac{37}{100}$ **d** $\dfrac{2}{100}$ **e** $\dfrac{85}{100}$

7 A sweatshirt costs £18.
   It is reduced by 20% in a sale.
   How much do you save if you buy the sweatshirt?

1 Draw a square.
   a Colour 50% of the square red.
   b Colour 25% of the square blue.
   c What percentage of the square is not coloured?

2 A man wins £120 000 on the lottery.
   He gives 10% to charity.
   a How much does he give to charity?
   b How much does he have left?

3 The percentage of males in the crowd watching a football match was 65%.
   The attendance at the match was 25 000.
   a How many males watched the match?
   b How many females watched the match?

4 Rachel carried out a survey of 150 patients at a hospital.
   She found out that 38% of the patients were children,
   26% were women and the rest were men.
   How many of the patients were:
   a children   **b** women   **c** men?

5 Martin is buying a Hi-Fi system.
   It costs £150.
   One shop offers a £25 discount.
   A second shop offers 15% off.
   Which should Martin buy?
   Show all your working.

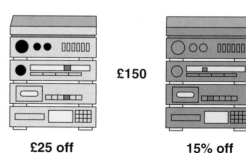

£150

£25 off   15% off

6 Jean got 70% in her maths exam.
   The exam was out of 90.
   How many marks did Jean get?

7 The full price of a coat is £80.
   It is reduced in a sale by 20%.
   What is the sale price of the coat?

8 There are 250 members of a golf club.
   72% are male.
   How many female members are there?

- $50\% = \dfrac{1}{2}$ $\quad 25\% = \dfrac{1}{4}$ $\quad 75\% = \dfrac{3}{4}$ $\quad 10\% = \dfrac{1}{10}$ $\quad 20\% = \dfrac{1}{5}$

- 23% of this pie-chart is blue.

  The fraction that is blue is $\dfrac{23}{100}$.

  You can change $\dfrac{23}{100}$ to a decimal.

  $\dfrac{23}{100} = 23 \div 100 = 0.23$

  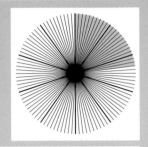

- 25% is the same as $\dfrac{1}{4}$.

  You divide by 4 to find 25%.

  Example Lisa is buying a tent for £120.
  Find how much Lisa saves.

  $\dfrac{1}{4}$ of £120 = £120 ÷ 4

  $\qquad\qquad\quad = £30$

  Lisa saves £30.

- You may need to use a calculator to work out a percentage.

  Example Sian is buying a computer.
  It costs £960.
  The shop reduces the price by 12%.
  How much does Sian save?

  Sian needs to work out 12% of 960.

	Step 1	**Step 2**	**Step 3**	**Step 4**
This can be written:	$\dfrac{12}{100}$	×	960	=

  Step 1 This changes the percentage to a decimal.
  The decimal appears when you press ☒ at step 2.

  Step 2 'Of' is the same as multiply so use the ☒ key.

  Step 3 You are finding the percentage of this amount.

  Step 4 This gives you the answer.

  Keys to press: **Step 1**   ¦ **Step 2** ¦ **Step 3**   ¦ **Step 4**

  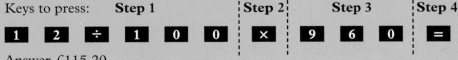

  Answer £115.20

1 For each diagram write down:
  a the percentage that is coloured,
  b the percentage that is not coloured.

(1) 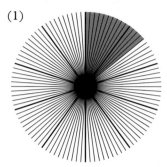   (2)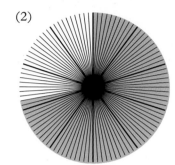

2 Gary says that 33% of his form went on a school trip.
What percentage did not go on the trip?

3 Copy these and fill in the missing numbers.

  a $10\% = \dfrac{...}{100} = \dfrac{1}{...}$

  c $75\% = \dfrac{...}{100} = \dfrac{3}{...}$

  b $20\% = \dfrac{...}{100} = \dfrac{1}{...}$

  d $70\% = \dfrac{...}{100} = \dfrac{...}{10}$

4 **a** What fraction of the
      counters are red?
  b What percentage of the
      counters are red?
  c What fraction of the counters are blue?
  d What percentage of the counters are blue?

5 Write these percentages as decimals.
  a 3%   **b** 34%   **c** 7%   **d** 98%

6 There were 60 pupils in the gym club last year.
This year the number went up by 15%.
  a How many more pupils go to gym club this year?
  b How many pupils go to gym club this year?

17 Improve your co-ordination

QUESTIONS

EXTENSION

SUMMARY

TEST YOURSELF

You plot co-ordinates using x and y axes.

Computers can plot graphs using 3 dimensions.

This is a graph of $z = \sin xy$.

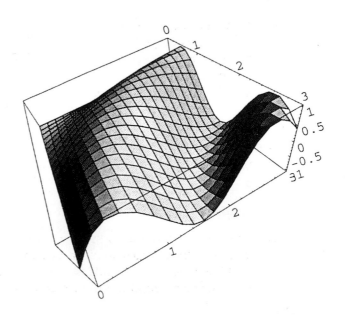

1 Co-ordinates

We use co-ordinates to find the position of a point.

Axes

We draw two lines. These are called **axes**.

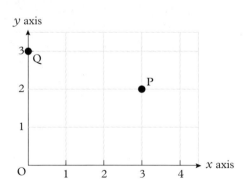

The co-ordinates of the point P are (3, 2).

x co-ordinate

The first number is the **x co-ordinate**.
The x co-ordinate of P is 3.
It tells you to move 3 squares *across*.

y co-ordinate

The second number is the **y co-ordinate**.
The y co-ordinate of P is 2.
It tells you to move 2 squares *up*.

The point Q has co-ordinates (0, 3).

Exercise 17:1

This picture shows the planets orbiting around the sun.

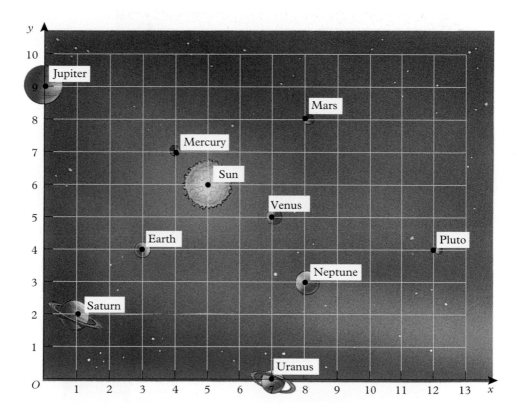

1 Write down the planet with co-ordinates:
  a (3, 4)   **c** (12, 4)
  b (7, 5)   **d** (4, 7)

2 Write down the co-ordinates of these planets:
  a Earth   **d** Saturn
  b Mars   **e** Uranus
  c Jupiter   **f** Neptune

Exercise 17:2

This is a map of the surface of the moon in 2100 AD.

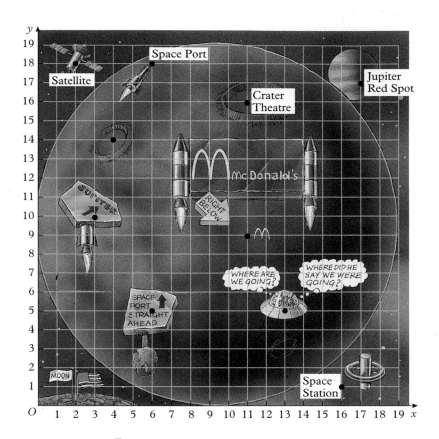

1 Write down what is at the point with co-ordinates:
   a (2, 18)   **c** (6, 5)
   b (13, 5)   **d** (3, 10)

2 Write down the co-ordinates of these places:
   a The top of the flag on the moon   **d** Space Port
   b Crater Theatre   **e** Space Station
   c Jupiter Red Spot   **f** McDonalds

Exercise 17:3

1 **a** Copy this set of axes on to
squared paper.

   b Plot the points.
Join them in order as you go.

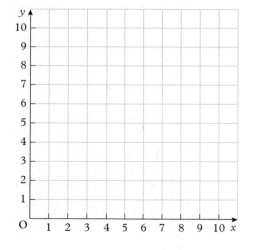

$(5, 4) \rightarrow (4, 4) \rightarrow (4, 1)$

$\rightarrow (3, 1) \rightarrow (2, 2) \rightarrow (3, 2)$

$\rightarrow (3, 6) \rightarrow (2, 6) \rightarrow (2, 5)$

$\rightarrow (1, 5) \rightarrow (1, 7) \rightarrow (4, 7)$

$\rightarrow (4, 8) \rightarrow (3, 8) \rightarrow (3, 10)$

$\rightarrow (5, 10)$

   c Join the point (5, 10) to the point
(5, 0) with a dotted line.
Reflect your shape in the dotted
line.

2 **a** Draw another set of axes like the ones in question **1**.

   b Plot the points.
Join them in order as you go.

$(5, 3) \rightarrow (3, 3) \rightarrow (1, 1)$

$\rightarrow (5, 1) \rightarrow (3, 3) \rightarrow (3, 5)$

$\rightarrow (2, 6) \rightarrow (3, 7) \rightarrow (5, 7)$

$\rightarrow (4, 8) \rightarrow (2, 8) \rightarrow (2, 10)$

$\rightarrow (4, 10) \rightarrow (4, 8)$

   c Put a mark like this ● at the point with co-ordinates (5, 10).

   d Join the point (5, 10) to the point (5, 0) with a dotted line.
Reflect your shape in the dotted line.

3 Jupiter is the largest planet.
This is a code.
Each pair of co-ordinates will give you a letter.
The code tells you one fact about Jupiter.

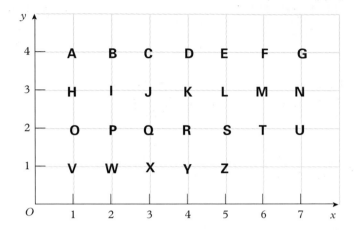

(3, 3) (7, 2) (2, 2) (2, 3) (6, 2) (5, 4) (4, 2) (2, 3) (5, 2)

(6, 3) (1, 2) (4, 2) (5, 4) (6, 2) (1, 3) (1, 4) (7, 3)

(1, 2) (7, 3) (5, 4)

(6, 2) (1, 3) (1, 2) (7, 2) (5, 2) (1, 4) (7, 3) (4, 4)

(6, 2) (2, 3) (6, 3) (5, 4) (5, 2)

(2, 4) (2, 3) (7, 4) (7, 4) (5, 4) (4, 2) (6, 2) (1, 3) (1, 4) (7, 3)

(5, 4) (1, 4) (4, 2) (6, 2) (1, 3)

4 Gavin and Petra are playing a game with dice.
They are using this set of axes.
They take it in turns to throw the two dice.
The red dice gives the *x* co-ordinate.
The blue dice gives the *y* co-ordinate.
The score is the number written on the axes
with these co-ordinates.

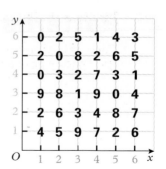

y						
6	0	2	5	1	4	3
5	2	0	8	2	6	5
4	0	3	2	7	3	1
3	9	8	1	9	0	4
2	2	6	3	4	8	7
1	4	5	9	7	2	6
O	1	2	3	4	5	6 x

Gavin rolls the dice.
The co-ordinates are (3, 5).
He scores 8.

Petra rolls the dice.
The co-ordinates are (4, 1).
She scores 7.

a These are all the co-ordinates that Gavin rolls:
   (3, 5) (2, 1) (4, 4) (6, 1)
   Find out all his scores.
   Add them together.
   What is Gavin's total score?

b These are all the co-ordinates that Petra rolls:
   (4, 1) (3, 2) (5, 6) (2, 2)
   Find out all her scores.
   Add them together.
   What is Petra's total score?

c The winner is the person with the highest score.
   Who is the winner?

d They decide to play a second game.
   Here are Gavin's co-ordinates: (3, 3) (6, 4) (3, 1) (1, 2)
   Here are Petra's co-ordinates: (6, 5) (2, 3) (1, 6) (4, 3)
   (1) What is Gavin's total score?
   (2) What is Petra's total score?
   (3) Who is the winner this time?

e Play the game with a friend.
   Roll the dice 4 times each.
   Work out your score to see who wins.

2 Patterns in co-ordinates

Peter has written the co-ordinates of a point on each of these cards.
He uses a rule to give some cards to Mary and some to John.

Exercise 17:4

1 a These are Mary's cards:
Look at the co-ordinates.
What do you notice?

(2, 4)	(5, 10)
(7, 14)	(3, 6)

b These are John's cards:
Look at the co-ordinates.
What do you notice?

(2, 3)	(10,11)
(8, 9)	(6, 7)

c Which of these cards
follow Mary's rule?

(3, 5)   (4, 8)   (1, 3)

d Which of these cards
follow John's rule?

(4, 5)   (7, 6)   (2, 7)

e One of these cards
follows Mary's rule
and John's rule.
Which card is it?

(3, 4)   (6, 3)   (1, 2)

Example   Look for a pattern in these co-ordinates.
Use the pattern to find the missing co-ordinates.

(2, 1) (3, 2) (4, 4) (5, 8) (..., ...), (..., ...)

These are the x co-ordinates:

2, 3, 4, 5   You add one each time.
The next two numbers are 6 and 7.

These are the y co-ordinates:

1, 2, 4, 8   You multiply by two each time.
The next two numbers are 16 and 32.

The missing co-ordinates are (6, 16) and (7, 32).

Look for a pattern in these co-ordinates.
Use the pattern to find the missing co-ordinates.

2   (2, 3) (4, 4) (6, 5) (8, 6) (..., ...) (..., ...)

3   (5, 6) (6, 7) (7, 8) (8, 9) (..., ...) (..., ...)

4   (3, 0) (5, 2) (7, 4) (9, 6) (..., ...) (..., ...)

5   (6, 9) (8, 8) (10, 7) (12, 6) (..., ...) (..., ...)

6   (5, 3) (10, 6) (15, 9) (20, 12) (..., ...) (..., ...)

Sometimes we need to look at the co-ordinates together to find a pattern.

Example

Look at these co-ordinates: (3, 5) (8, 0) (2, 6) (4, 4)
  a Describe the pattern.
  b Use your pattern to fill in the missing numbers of these co-ordinates: (0, ...) (7, ...) (5, ...)

  a The two numbers add up to eight.
    $3 + 5 = 8$ $8 + 0 = 8$
    $2 + 6 = 8$ $4 + 4 = 8$
  b The missing numbers are (0, 8) (7, 1) (5, 3).

Look for a pattern in these co-ordinates.
Use the pattern to find the missing co-ordinates.

7 (1, 9) (6, 4) (5, 5) (10, 0) (3, ...) (..., 2)

8 (5, 1) (6, 0) (2, 4) (4, ...) (..., 3) (0, ...)

9 (3, 12) (5, 10) (7, 8) (2, 13) (0, ...) (..., 9)

10 (6, 14) (12, 8) (17, 3) (11, 9) (..., 1) (4, ...)

11 (5, 3) (2, 6) (4, 4) (9, −1) (1, ...) (..., −2)

Exercise 17:5

1 Luke joins four points on
a grid to make a square.
The points are:
(0, 0) (2, 0) (2, 2) (0, 2)
The area of Luke's square
is 4 cm².

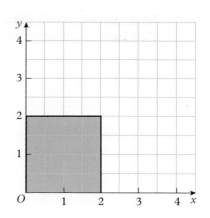

Karen multiplies each of
Luke's co-ordinates by 2.

Luke's co-ordinates	×2	Karen's co-ordinates
(0, 0)		(0, 0)
(2, 0)		(4, 0)
(2, 2)		(4, 4)
(0, 2)		(0, 4)

a Copy the axes.
b Draw Luke's square.
c Draw Karen's square.
d Find the area of Karen's square.

2 Jenny multiplies each of Luke's co-ordinates by 3.
a Copy and complete:

Luke's co-ordinates	×3	Jenny's co-ordinates
(0, 0)		(..., ...)
(2, 0)		(..., ...)
(2, 2)		(..., ...)
(0, 2)		(..., ...)

b Plot Jenny's co-ordinates.
c Draw Jenny's square.
d Find the area of Jenny's square.

3 Alan multiplies Luke's co-ordinates by another number.
Two of Alan's points are (12, 0) and (0, 12).
a What number did Alan use?
b What are the other two points?
c What will be the area of Alan's square?

4 Chris has some red tiles and some blue tiles.
Chris places some red tiles in a row like this:

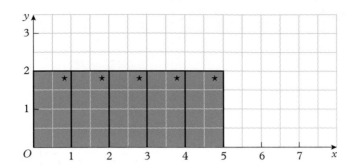

Each tile has a corner marked with a ⋆.
The co-ordinates of the first corner are (1, 2).

a Write down the co-ordinates of the first five corners that have a ⋆.

b Look at the numbers in the co-ordinates.
Describe two things that you notice.

c Chris places some more red tiles in the row.
Write down the co-ordinates of the ⋆ on the sixth tile.

d Chris thinks that (12, 3) are the co-ordinates of one of the corners
with a ⋆.
Explain why he is wrong.

Chris now places some blue tiles above the red tiles like this:

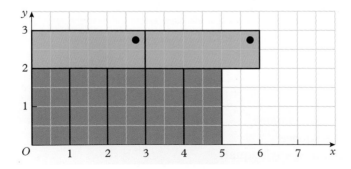

e Write down the co-ordinates of the first three corners that have a ●.

f Look at the numbers in the co-ordinates.
Describe two things that you notice.

g What will be the co-ordinates of the tenth corner that has a ●?

5 Three points on this line are marked with a ✕.
Their co-ordinates are:
(0, 2) (1, 3) (3, 5).

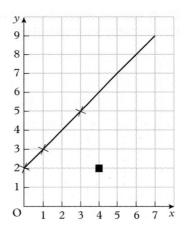

a Look at the numbers in the co-ordinates of each point.
What do you notice?

b These points lie on the line.
Copy them and fill in the missing numbers.
(2, …)   (…, 7)   (12, …)

c The point ■ is below the line.
Write down its co-ordinates.

d Four points are at (3, 6) (5, 8) (7, 8) (9, 11).
Which one of these points is below the line?
Explain how you know from the co-ordinates.

e The point (…, 3) is above the line.
Copy and fill in a possible co-ordinate for the point.

1   Copy this set of axes:

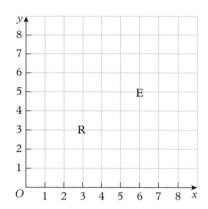

Write each letter at the point given by the co-ordinates.
The first two are done for you.

E (6, 5)   R (3, 3)   O (1, 1)   P (3, 7)

L (7, 2)   M (5, 1)   H (1, 6)   S (3, 2)

N (8, 5)   O (2, 4)   E (6, 7)   O (7, 1)   H (2, 5)   U (2, 7)   Y (8, 2)

T (5, 5)   S (5, 4)   E (3, 1)   T (5, 7)   R (4, 5)   O (5, 2)   N (8, 1)

S (3, 6)   E (1, 3)   A (2, 2)   H (5, 3)   I (4, 7)   O (6, 1)   T (1, 5)

N (4, 4)   E (7, 5)   R (7, 7)   H (1, 2)   J (1, 7)   I (3, 5)   M (1, 4)

N (2, 1)   T (4, 3)   A (2, 3)   N (6, 2)   O (3, 4)   A (2, 6)

The letters will spell out a message.

2   Look for a pattern in these co-ordinates.
Use the pattern to find the missing co-ordinates.

   a   (5, 4) (6, 5) (7, 6) (8, 7) (…, …) (…, …)

   b   (3, 4) (6, 6) (9, 8) (12, 10) (…, …) (…, …)

   c   (5, 4) (6, 3) (0, 9) (1, 8) (…, 2) (9, …)

   d   (7, 4) (6, 5) (9, 2) (0, 11) (11, …) (…, 3)

3 This map shows some of the villages close to Swindon.
Write down the names and co-ordinates of the villages that are:
a due north of Swindon
b due east of Swindon
c south west of Swindon
d north east of Swindon

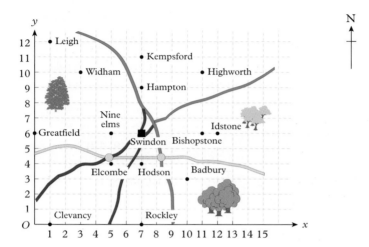

4 **a** Copy these axes on to squared paper.
b Plot these co-ordinates:
(0, 0) (2, 1) (6, 3) (8, 4)
c Find the y co-ordinate of (4, ...) to complete the pattern.
d Join the points in a straight line with a ruler.
e Which of these points is not above the line?
Explain how you know from the co-ordinates.
(3, 4) (7, 7) (5, 1) (1, 1)

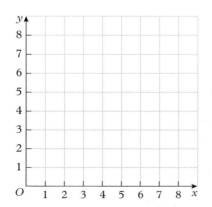

1 Look at these two sets of cards:

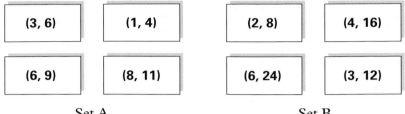

| (3, 6) | (1, 4) | (2, 8) | (4, 16) |
| (6, 9) | (8, 11) | (6, 24) | (3, 12) |

Set A   Set B

  a What do you notice about the numbers on each card in set A?
  b What do you notice about the numbers on each card in set B?
  c Which set do these cards belong to?

(10, 13)   (10, 40)   (5, 20)   (9, 12)

  d This card can belong to either set.
     Can you find the missing numbers?   (. . ., . . .)

  e This is another set of cards.
     What are the missing numbers?

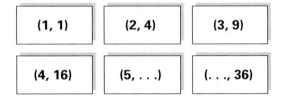

| (1, 1) | (2, 4) | (3, 9) |
| (4, 16) | (5, . . .) | (. . ., 36) |

2 Look for a pattern in these co-ordinates.
  Use the pattern to find the missing co-ordinates.
  a (18, 4) (16, 5) (14, 7) (12, 10) (…, …) (…, …)
  b (2, 1) (4, 9) (8, 17) (14, 25) (…, …) (…, …)
  c (1, 80) (4, 40) (9, 20) (16, 10) (…, …) (36, …)

- **Axes**

We draw two lines.
These are called
axes.

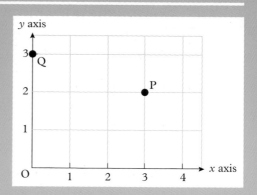

The co-ordinates of the point P are (3, 2).

- **x co-ordinate**

The first number is the **x co-ordinate**.
The *x* co-ordinate of P is 3.
It tells you to move 3 squares *across*.

- **y co-ordinate**

The second number is the **y co-ordinate**.
The *y* co-ordinate of P is 2.
It tells you to move 2 squares *up*.

The point Q has co-ordinates (0, 3).

- *Example*

Look for a pattern in these co-ordinates.
Use the pattern to find the missing co-ordinates:

(2, 1) (3, 2) (4, 4) (5, 8) (…, …), (…, …)

These are the *x* co-ordinates:

2, 3, 4, 5   You add one each time.
              The next two numbers are **6** and **7**.

These are the *y* co-ordinates:

1, 2, 4, 8   You multiply by two each time.
              The next two numbers are **16** and **32**.

The missing co-ordinates are (**6**, **16**) and (**7**, **32**)

1 **a** Copy this set of axes.
    b Plot these points.
        Join them up with a ruler as you go.

        $(5, 10) \rightarrow (4, 9) \rightarrow (4, 6) \rightarrow (1, 3)$
        $\rightarrow (4, 3) \rightarrow (3, 1) \rightarrow (5, 1)$

    c Join the point (5, 10) to the point
        (5, 0) with a dotted line.
    d Reflect your shape in the line.

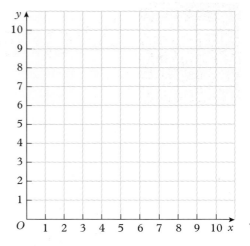

2 Rachid has some tiles in the shape of a parallelogram.
    He has placed them like this.
    Each tile has a corner marked with a *.
    The co-ordinates of the first corner are (3, 2).

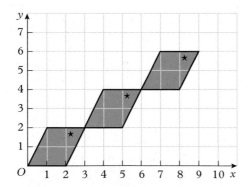

    a Write down the co-ordinates of the first three corners that have a *.
    b Look at the numbers in the co-ordinates.
        Describe two things that you notice.
    c What will be the co-ordinates of the fourth and fifth corners that
        have a *?
    d Rachid thinks that (21, 17) are the co-ordinates of one of the
        corners with a *.
        Explain why he is wrong.

18 More fun

1 Fruit machine

This fruit machine has three reels.
Each reel has pictures of fruit on it.

You are going to design your own fruit machine.
Here is a simple one to start you off.
You will use strips for the reels.
It has just two reels.
Each reel has three fruits.
Here are the fruits that are on each reel.

The only way to win on this machine is to get two plums.

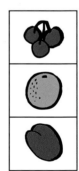

1 You need worksheets 18:1 and 18:2.
Cut out the strips from worksheet 18:1.
Cut out the slots from worksheet 18:2.
Fit the strips into the first two reels of the fruit machine.

2 Start with 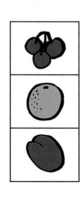 in both windows.

You are now going to list all the different ways the machine could stop.

3   Copy this table:

Reel 1	Reel 2
Cherries	Cherries
Cherries	
Cherries	

You need 6 more rows.

4   Move reel 2 one space up.
Fill in the next row of your table.

5   Move reel 2 again.
Fill in the next row of your table.

6   Move reel 1 one space up.

Put reel 2 back to .

Fill in the next row of your table.

7   Carry on until you have filled in the rest of your table.

8   How many different ways can the machine stop?

9   What is the probability of winning?

W 10   Now design your own fruit machines. You could:

- Have more fruits.
- Use three reels.
- Have more prizes.
- Put the same fruit on a reel more than once.

For each machine that you make:

- Work out how many ways the machine can stop.
- Work out the probability of winning.

2 Off we go!

You are going to plan a weekend in Paris.
This is what you'll need to do:

- Leave any time after school on Friday.
- Be back by midnight on Sunday.

- Decide how you want to get there (and back!).
- Plan your journey times carefully.

- Decide how many people are going with you.
- Work out how much it costs.

You might decide to go by train.
This means going through the Channel Tunnel.
Here are some questions for you to think about.

- What time is your train to London?
  (use worksheet 18:7)
- Which London station will you get to?
- How long will you allow to get to Waterloo?
- What time does the Eurostar train leave?
  (use worksheet 18:8)
- What time will you get to Paris?
- What will you do in Paris?
  (use worksheets 18:9, 18:10)
- What time will you leave Paris?
- What time will you get home?
- How much will it cost?

You might go by car and ferry.

- How long will it take to get to the ferry?
- How long will the ferry take?
  (use worksheet 18:11)
- How long will the drive to Paris take?
  (use worksheet 18:12)
- What time will you get to Paris?
- What will you do in Paris?
  (use worksheets 18:9, 18:10)
- What time will you leave Paris?
- What time will you get home?
- How much will it cost?

3 Planning a bedroom

W You are going to design the layout of a
bedroom.
Use worksheets 18:13, 18:14, 18:15.

This is what you will need to do:

- Have a plan of the bedroom.
- Think about how much you can spend.
- Decide what you want to buy.
- Decide which range you want to buy.
- Cut out rectangles to represent each
  piece of furniture.
- Put the furniture where you want it
  in the room.

Make sure you don't block the door or
the window.

- When you are happy with the layout,
  stick the furniture in place.
- Work out the total cost of the furniture.

Help yourself

1 Adding

You should set out additions in columns.

Example

13 + 2   should be set out like this:

```
   13
 +  2
   15
```

Here are some more examples.

27 + 21	431 + 26	542 + 136
27	431	542
+ 21	+ 26	+ 136
48	457	678

Sometimes you need to 'carry'. This happens when a column adds up to 10 or more.

Example

13 + 9

```
   13
 +  9
   22
    ₁
```

Here are some more examples.

27 + 29	246 + 28	558 + 67
27	246	558
+ 29	+ 28	+ 67
56	274	625
₁	₁	₁ ₁

Exercise 1

Copy these into your book.
Work out the answers.

1
```
   46
 +  3
```

2
```
   54
 + 32
```

3
```
   86
 + 12
```

4
```
  762
 + 134
```

5
```
  401
 +  48
```

6
```
   35
 + 853
```

7
```
   6214
 +  735
```

8
```
   2418
 + 6300
```

9 31 + 8

10 64 + 35

11 615 + 31

12 814 + 154

Exercise 2

Copy these into your book.
Work out the answers.

1
```
   32
 +  9
```

2
```
   44
 + 18
```

3
```
   38
 + 27
```

4
```
  147
 +  35
```

5
```
  285
 + 16
```

6
```
  449
 + 82
```

7
```
  268
 + 173
```

8
```
  629
 + 216
```

9 38 + 46

10 95 + 54

11 114 + 99

12 57 + 333

Other words

All these words can also mean **add**.

   plus   **sum**   **total**

Examples
Work out 24 **plus** 13
Find the **sum** of 24 and 13
Find the **total** of 24 and 13

$$\begin{array}{r} 24 \\ +\ 13 \\ \hline 37 \end{array}$$

all mean

2 Subtracting

Subtractions should also be set out in columns.

Example

28 − 10   should be set out like this:

$$\begin{array}{r} 28 \\ -\ 10 \\ \hline 18 \end{array}$$

Here are some more examples.

29 − 16   436 − 25   587 − 226

$$\begin{array}{r} 29 \\ -\ 16 \\ \hline 13 \end{array} \qquad \begin{array}{r} 436 \\ -\ 25 \\ \hline 411 \end{array} \qquad \begin{array}{r} 587 \\ -\ 226 \\ \hline 361 \end{array}$$

Exercise 3

Copy these into your book.
Work out the answers.

1
$$\begin{array}{r} 85 \\ -\ 23 \\ \hline \end{array}$$

3
$$\begin{array}{r} 345 \\ -\ 123 \\ \hline \end{array}$$

5
$$\begin{array}{r} 398 \\ -\ 102 \\ \hline \end{array}$$

2
$$\begin{array}{r} 756 \\ -\ 245 \\ \hline \end{array}$$

4
$$\begin{array}{r} 575 \\ -\ 42 \\ \hline \end{array}$$

6
$$\begin{array}{r} 1928 \\ -\ 416 \\ \hline \end{array}$$

7
$$\begin{array}{r} 8827 \\ -\ 714 \\ \hline \end{array}$$

10 873 − 352

8
$$\begin{array}{r} 3924 \\ -\ 600 \\ \hline \end{array}$$

11 543 − 140

9 659 − 47

12 3579 − 2254

Sometimes you need to 'borrow'. This happens when the number on the bottom of a column is bigger than the one on the top.

Example

42 − 19

The 4 is worth 4 lots of 10.
You can 'borrow' one of these 10s.
You change it into ten ones.

$$\begin{array}{r} 42 \\ -\ 19 \\ \hline \end{array} \leftarrow \text{The 9 is bigger than the 2.}$$

Your working now looks like this:

$$\begin{array}{r} {}^{3}\cancel{4}{}^{1}2 \\ -\ 19 \\ \hline 23 \end{array} \leftarrow \text{You can now take the 9 away from the 12}$$

Here is another example:

64 − 28

$$\begin{array}{r} 64 \\ -\ 28 \\ \hline \end{array} \rightarrow \begin{array}{r} {}^{5}\cancel{6}{}^{1}4 \\ -\ 28 \\ \hline \end{array} \rightarrow \begin{array}{r} {}^{5}\cancel{6}{}^{1}4 \\ -\ 28 \\ \hline 36 \end{array}$$

Here are some more difficult examples:

82 − 67   231 − 119   623 − 487

$$\begin{array}{r} {}^{7}\cancel{8}{}^{1}2 \\ -\ 67 \\ \hline 15 \end{array} \qquad \begin{array}{r} 2{}^{2}\cancel{3}{}^{1}1 \\ -\ 119 \\ \hline 112 \end{array} \qquad \begin{array}{r} {}^{5}\cancel{6}{}^{11}\cancel{2}{}^{1}3 \\ -\ 487 \\ \hline 136 \end{array}$$

357

Exercise 4

Copy these into your book.
Work out the answers.

1
```
    36
  − 18
  ────
```

2
```
    95
  − 49
  ────
```

3
```
    65
  − 38
  ────
```

4
```
   243
  −  25
  ─────
```

5
```
   532
  − 428
  ─────
```

6
```
   725
  − 156
  ─────
```

7
```
   193
  −  44
  ─────
```

8
```
   342
  − 138
  ─────
```

9 624 − 255

10 621 − 147

11 726 − 508

12 526 − 347

You cannot borrow from the next
column if there is a zero in it.
You may need to borrow across more
than one column.

Example

$$
\begin{array}{r} 300 \\ -\ 196 \\ \hline \end{array}
\rightarrow
\begin{array}{r} {\scriptstyle 2}\,{\scriptstyle 1}\!3\!\!\!/\,00 \\ -\ 196 \\ \hline \end{array}
\rightarrow
\begin{array}{r} {\scriptstyle 2}\,{\scriptstyle 9}\,{\scriptstyle 1}\!3\!\!\!/\,\!0\!\!\!/\,0 \\ -\ 196 \\ \hline 104 \end{array}
$$

Exercise 5

Copy these into your book.
Work out the answers.

1
```
   600
  − 173
  ─────
```

2
```
   700
  − 347
  ─────
```

3
```
   8000
  −  285
  ──────
```

4
```
   7000
  − 5416
  ──────
```

Other words

All these words can also mean
subtract.

> **take away**   **take**
>
> **minus**   **difference**

Examples
Find 73 **take away** 24
Work out 73 **take** 24
Find 73 **minus** 24
Find the **difference** between
73 and 24

Checking

You can always check a subtraction
by adding.

Example

256 − 183

$$
\begin{array}{r} {\scriptstyle 1}\,{\scriptstyle 1}\\ 2\!\!\!/\,56 \\ -\ 183 \\ \hline 73 \end{array}
\qquad \text{check} \qquad
\begin{array}{r} 183 \\ +\ \ 73 \\ \hline 256 \end{array}
$$

Go back to your answers for Exercise **5**.
Check each of them by adding.

3 Multiplying

When you are adding lots of the same
number it is quicker to multiply.

Example

$$
\begin{array}{r} 31 \\ 31 \\ 31 \\ 31 \\ +\ \ 31 \\ \hline 155 \end{array}
\quad \text{is the same as} \quad
\begin{array}{r} 31 \\ \times\ \ 5 \\ \hline 155 \end{array}
$$

To do $\quad\begin{array}{r}31\\ \times\quad5\\ \hline\end{array}$ first do $\quad\begin{array}{r}31\\ \times\quad5\\ \hline 5\end{array}$ 5×1

then do 5×3 $\quad\begin{array}{r}31\\ \times\quad5\\ \hline 155\end{array}$

Remember to keep your numbers in columns.

Here are some more examples:

$\begin{array}{r}62\\ \times\quad4\\ \hline 248\end{array}\qquad\begin{array}{r}51\\ \times\quad9\\ \hline 459\end{array}$

Exercise 6

1 $\begin{array}{r}33\\ \times\quad2\\ \hline\end{array}$ **3** $\begin{array}{r}31\\ \times\quad4\\ \hline\end{array}$

2 $\begin{array}{r}132\\ \times\quad3\\ \hline\end{array}$ **4** $\begin{array}{r}133\\ \times\quad3\\ \hline\end{array}$

Sometimes you need to carry.

Example

$\begin{array}{r}26\\ \times\quad3\\ \hline 8\\ {\scriptstyle 1}\end{array}\quad\rightarrow\quad\begin{array}{r}26\\ \times\quad3\\ \hline 78\\ {\scriptstyle 1}\end{array}$

$3\times2=6$
Then add the 1 to give 7

Exercise 7

1 $\begin{array}{r}45\\ \times\quad2\\ \hline\end{array}$ **3** $\begin{array}{r}54\\ \times\quad4\\ \hline\end{array}$ **5** $\begin{array}{r}349\\ \times\quad2\\ \hline\end{array}$

2 $\begin{array}{r}35\\ \times\quad2\\ \hline\end{array}$ **4** $\begin{array}{r}125\\ \times\quad3\\ \hline\end{array}$ **6** $\begin{array}{r}428\\ \times\quad5\\ \hline\end{array}$

---

7 $\begin{array}{r}157\\ \times\quad9\\ \hline\end{array}$ **9** $\begin{array}{r}555\\ \times\quad5\\ \hline\end{array}$

8 $\begin{array}{r}634\\ \times\quad9\\ \hline\end{array}$ **10** $\begin{array}{r}901\\ \times\quad4\\ \hline\end{array}$

Other words

These words can also mean **multiply**.

times   **product**   **of**

Examples
Find 24 **times** 16
Find the **product** of 24 and 16
Find one half **of** 24

---

4 Multiplying by 10

When you multiply by 10, all the digits move across **one** column to the **left**. This makes the number 10 times bigger.
You can use the headings **Th H T U** to help.
They mean **Th**ousands, **H**undreds, **T**ens and **U**nits. Units is another way of saying 'ones'.

Example
$23\times10=230$

H   T   U

$\begin{array}{ccc}&2&3\\2&3&0\end{array}$
$\times10 \quad \times10$

Here are some more examples:

Th H T U

$\overset{4}{}\overset{6}{}$ $46 \times 10 = 460$
4   6   0

$\overset{2}{}\overset{5}{}\overset{3}{}$ $253 \times 10 = 2530$
2   5   3   0

$\overset{6}{}\overset{0}{}\overset{1}{}$ $601 \times 10 = 6010$
6   0   1   0

Exercise 8

Multiply these numbers by 10

1 48   **4** 842   **7** 7000
2 54   **5** 777   **8** 9003
3 134   **6** 2496

5 Multiplying by 100, 1000, ...

When you multiply by 100, all the digits move across **two** columns to the **left**.
This makes the number 100 times bigger.
This is because $100 = 10 \times 10$
So multiplying by 100 is like multiplying by 10 twice.

Example

$74 \times 100 = 7400$
Th H T U

7   4   0   0

When you multiply by 1000 all the numbers move across three columns to the left.

This is because $1000 = 10 \times 10 \times 10$
This means that multiplying by 1000 is like multiplying by 10 three times.

Example

$74 \times 1000 = 74\,000$

TTh Th H T U

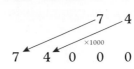
7   4   0   0   0

Exercise 9

Write down the answers to these.

1 27×100 **7** 4153×100

2 91×100 **8** 900×1000

3 74×1000 **9** 4004×1000

4 291×100 **10** $924 \times 10\,000$

5 4270×100 **11** $301 \times 10\,000$

6 840×1000 **12** $737 \times 100\,000$

6 Multiplying by 20, 30, ...

When you multiply by 20 it is like multiplying by 2 then by 10. This is because $20 = 2 \times 10$

Example

To do 18×20
first do   1 8
                   $\times$ 2
                   ———
                     3 6
                      1

Then do $36 \times 10 = 360$

So $18 \times 20 = 360$

In the same way multiplying by 30 is the same as multiplying by 3 and then multiplying by 10

Example

To do 26×30:

first do
$$
\begin{array}{r}
26 \\
\times \quad 3 \\
\hline
78 \\
\end{array}
$$

Then do $78 \times 10 = 780$

So $26 \times 30 = 780$

Exercise 10

Work these out.

1 39×20

2 42×20

3 26×30

4 23×30

5 65×30

6 34×40

7 92×40

8 25×50

9 71×50

10 304×20

11 291×30

12 525×70

7 Long multiplication

When you want to multiply two quite large numbers you have to do it in stages. Here are two methods.
You only have to know one of them.

Method 1

Example 146×24

First do 146×4
$$
\begin{array}{r}
146 \\
\times \quad 4 \\
\hline
584 \\
\end{array}
$$

Then do 146×20
$$
\begin{array}{r}
146 \\
\times \quad 2 \\
\hline
292 \\
\end{array}
$$

$292 \times 10 = 2920$

Now add the two answers together.
$$
\begin{array}{r}
584 \\
+ \quad 2920 \\
\hline
3504 \\
\end{array}
$$

Usually the working out looks like this:
$$
\begin{array}{r}
146 \\
\times \quad 24 \\
\hline
584 \\
2920 \\
\hline
3504 \\
\end{array}
$$

Here is another example.
$$
\begin{array}{r}
223 \\
\times \quad 36 \\
\hline
1338 \\
6690 \\
\hline
8028 \\
\end{array}
$$
$\leftarrow (223 \times 6)$
$\leftarrow (223 \times 30)$

361

Method 2

Example 125×23

First set out the numbers with boxes,
like this:

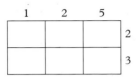

Now draw in the diagonals like this:

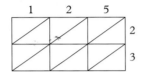

Fill in like a table square then add
along the diagonals like this:

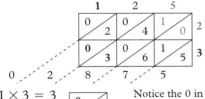

$1 \times 3 = 3$

Notice the 0 in the
top box when the
answer is a single
digit.

So $125 \times 23 = \mathbf{2875}$

Here is another example.
When the diagonal adds up to more
than 10, you carry into the next one.

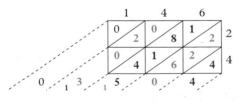

So $146 \times 24 = \mathbf{3504}$

Exercise 11

Use the method you prefer to work
these out.

1	27×25	**7**	391×45
2	76×24	**8**	317×84
3	123×53	**9**	545×22
4	404×26	**10**	821×65
5	382×25	**11**	754×71
6	271×54	**12**	989×89

8 Dividing

Multiplying is like doing lots of
additions. In the same way dividing is
like doing lots of subtractions.

To find out how many 4's make 12
you can see how many times you can
take 4 away from 12

$12 - 4 = 8$ (once)
$\quad 8 - 4 = 4$ (twice)
$\quad 4 - 4 = 0$ (three times)

So there are 3 lots of 4 in 12

You can say 12 divided by 4 is 3

or $12 \div 4 = 3$

Example
$\qquad 15 \div 3 = ?$
$\qquad 15 - 3 = 12$ (once)
$\qquad 12 - 3 = 9$ (twice)
$\qquad\quad 9 - 3 = 6$ (three times)
$\qquad\quad 6 - 3 = 3$ (four times)
$\qquad\quad 3 - 3 = 0$ (five times)

$\qquad$ So $15 \div 3 = 5$

Exercise 12

Work these out.

1	$15 \div 3$	**7**	$36 \div 9$
2	$55 \div 5$	**8**	$70 \div 10$
3	$24 \div 6$	**9**	$16 \div 2$
4	$16 \div 4$	**10**	$54 \div 6$
5	$36 \div 4$	**11**	$40 \div 5$
6	$28 \div 7$	**12**	$81 \div 9$

When the numbers get bigger, this method takes too long. You need a new way to work it out.

Example

$68 \div 2$

$2\overline{)68}$

First work out $6 \div 2 = 3$. Put the 3 above the 6:

$$2\overline{)\overset{3}{6}8}$$

Now work out $8 \div 2 = 4$. Put the 4 above the 8:

$$2\overline{)\overset{34}{68}}$$

So $68 \div 2 = 34$

Here is another example: $84 \div 4$

$$4\overline{)\overset{21}{84}}$$

So $84 \div 4 = 21$

Exercise 13

Work these out.

1	$2\overline{)66}$	**4**	$66 \div 3$
2	$3\overline{)39}$	**5**	$82 \div 2$
3	$8\overline{)88}$	**6**	$484 \div 4$

Sometimes you need to 'carry'. This happens when a number does not divide exactly.

Example

$72 \div 4$

$4\overline{)72}$

First do $7 \div 4$. This is 1 with 3 left over.
Put the 1 above the 7 and carry the 3 like this:

$$4\overline{)\overset{1}{7^{3}2}}$$

Now do $32 \div 4$. This is 8. Put the 8 above the 2 like this:

$$4\overline{)\overset{18}{7^{3}2}}$$

So $72 \div 4 = 18$

Here is another example: $75 \div 5$

$$5\overline{)\overset{15}{7^{2}5}}$$

So $75 \div 5 = 15$

Exercise 14

Work these out.

1 $2\overline{)54}$ **2** $3\overline{)51}$

363

3 64 ÷ 4

4 52 ÷ 4

5 91 ÷ 7

6 68 ÷ 4

7 90 ÷ 6

8 144 ÷ 8

9 432 ÷ 4

10 284 ÷ 2

11 531 ÷ 3

12 378 ÷ 7

9 Dividing by 10

When you divide by 10, all the digits move across **one** column to the **right**. This makes the number smaller.

Example

230 ÷ 10 = 23

Here are some more examples.

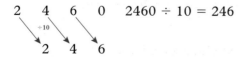

580 ÷ 10 = 58

2460 ÷ 10 = 246

Exercise 15

Divide these numbers by 10

1 820

2 60

3 4820

4 930

5 8160

6 9400

7 7000

8 500 000

364

10 Dividing by 100, 1000, ...

When you divide by 100, all the digits move across **two** columns to the **right**. This is because 100 = 10 × 10 So dividing by 100 is like dividing by 10 twice.

Example

7400 ÷ 100 = 74

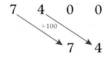

When you divide by 1000, all the numbers move across **three** columns to the **right**.

Example

74 000 ÷ 1000 = 74

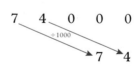

Exercise 16

Work these out.

1 5400 ÷ 100

2 7100 ÷ 100

3 8200 ÷ 100

4 64 000 ÷ 1000

5 84 000 ÷ 100

6 84 000 ÷ 1000

7 400 000 ÷ 1000

8 400 000 ÷ 10 000

11 Dividing by 20, 30, ...

When you divide by 20, it is like dividing by 2 then by 10. This is because $20 = 2 \times 10$

Example

To do $360 \div 20$

first do
$$\begin{array}{r} 1\,8\,0 \\ 2\overline{)3\,^{1}6\,0} \end{array}$$

Then do $\quad 180 \div 10 = 18$

So $\qquad 360 \div 20 = 18$

In the same way dividing by 30 is the same as dividing by 3 then by 10

Example

To do $780 \div 30$

first do
$$\begin{array}{r} 2\,6\,0 \\ 3\overline{)7\,^{1}8\,0} \end{array}$$

Then do $\quad 260 \div 10 = 26$

So $\qquad 780 \div 30 = 26$

Exercise 17

Work these out.

1 $820 \div 20$ **5** $7520 \div 20$

2 $480 \div 30$ **6** $4620 \div 30$

3 $3720 \div 40$ **7** $1980 \div 90$

4 $5250 \div 50$ **8** $24\,480 \div 80$

Other words

These words can also mean **divide**.

    share   **quotient**

Examples

Share 240 by 12
Find the **quotient** $\Big\}$ both mean
of 240 and 12 $\qquad$ $240 \div 12$

12 Subtracting fractions

Example

$$\frac{3}{5} - \frac{2}{5} = \frac{1}{5}$$

The two bottom numbers must still be the same.

Example

$$\frac{3}{8} - \frac{1}{4}$$

Numbers that 8 goes into:
⑧ 16 24 ...

Numbers that 4 goes into:
4 ⑧ 12 16 ...

$$\frac{1}{4} = \frac{?}{8} \qquad \overset{\times 2}{\frac{1}{4}} = \frac{2}{8}$$
$$\underset{\times 2}{} \qquad \underset{\times 2}{}$$

The $\dfrac{3}{8}$ does not need changing.

So $\dfrac{3}{8} - \dfrac{1}{4} = \dfrac{3}{8} - \dfrac{2}{8} = \dfrac{1}{8}$

Exercise 18

Work these out.

1 $\dfrac{5}{8} - \dfrac{2}{8}$ **3** $\dfrac{6}{13} - \dfrac{5}{13}$

2 $\dfrac{3}{5} - \dfrac{2}{5}$ **4** $\dfrac{2}{5} - \dfrac{1}{10}$

5 $\dfrac{6}{8} - \dfrac{1}{4}$ **8** $\dfrac{2}{4} - \dfrac{1}{3}$

6 $\dfrac{11}{12} - \dfrac{1}{3}$ **9** $\dfrac{7}{8} - \dfrac{2}{3}$

7 $\dfrac{1}{4} - \dfrac{1}{6}$ **10** $\dfrac{3}{4} - \dfrac{1}{3}$

13 Converting units

Common metric units of length

10 millimetres (mm) = 1 centimetre (cm)
100 centimetres = 1 metre (m)
1000 metres = 1 kilometre (km)

Examples

1 Convert 6.9 cm to mm.
6.9 cm = 6.9 × 10 mm
= 69 mm

2 Convert 5.34 m to cm.
5.34 m = 5.34 × 100 cm
= 5.34 cm

3 Convert 7.3 km to m.
7.3 km = 7.3 × 1000 m
= 7300 m

Exercise 19

1 Convert these lengths to mm.
  a 3.4 cm   **c** 131 cm
  b 12.8 cm   **d** 113.7 cm

2 Convert these lengths to cm.
  a 4.1 m   **c** 12.1 m
  b 2.8 m   **d** 324 m

3 Convert these lengths to m.
  a 8.8 km   **c** 15 km
  b 9.7 km   **d** 100 km

Common metric units of mass

Units of mass have similar names to units of length.

1000 milligrams (mg) = 1 gram (g)
1000 grams = 1 kilogram (kg)
1000 kg = 1 tonne (t)

Examples

1 Convert 2.5 kg to g.
2.5 kg = 2.5 × 1000 g
= 2500 g

2 Convert 5000 g to kg.
5000 g = 5000 ÷ 1000 kg
= 5 kg

Exercise 20

Convert the units in each of these.
Think carefully whether you need to
multiply or divide.

1 **a** 3 kg to g   **d** 3000 g to kg
    b 7 kg to g   **e** 6000 g to kg
    c 21 kg to g   **f** 800 g to kg

2 **a** 3.5 kg to g   **d** 6500 g to kg
    b 7.2 kg to g   **e** 3200 g to kg
    c 3.84 kg to g   **f** 2800 g to kg

3 **a** 5 g to mg   **d** 3000 kg to t
    b 8000 mg to g   **e** 4 t to kg
    c 1640 mg to g   **f** 2.5 t to kg

CHAPTER 1

1   **a** 4 cm^2 **b** 5 cm^2

2   **a** $6 \times 13 = 78 \text{ cm}^2$ **c** $7 \times 9 = 63 \text{ cm}^2$

    b $\dfrac{10 \times 12}{2} = 60 \text{ cm}^2$ **d** $\dfrac{10 \times 8}{2} = 40 \text{ cm}^2$

3   **a** $\dfrac{3 \times 3}{2} = 4.5 \text{ cm}^2, \dfrac{2 \times 3}{2} = 3 \text{ cm}^2$

    b $1 \times 4 = 4 \text{ cm}^2$

CHAPTER 2

1   **a** 14×7
    b   98

2   **a**   8 packs   **b**   8 times   **c**   3 cans

3   **a**   254   **b**   1

4   **a**   7   **b**   9   **c**   9   **d**   9

5   **a**   7   **b**   13   **c**   28   **d**   21   **e**   34

6   **a**   371   **b**   6072

7   **a**   67   **b**   206   **c**   54

8   28 remainder 2

CHAPTER 3

1   **a** 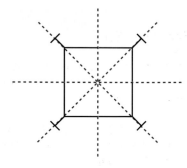   **b**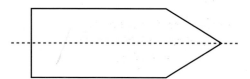

2   **a** 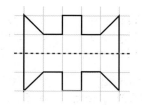   **b**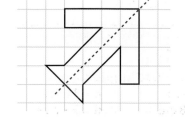

3 **a** 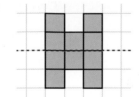 **b**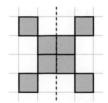

4 **a** 4, a square fits on top of itself 4 times as it makes a complete turn.
  b 2, the oval fits on top of itself twice as it makes a complete turn.

CHAPTER 4

1 **a** 5   **b** 2   **c** 7

2 **a** 12 babies   **c** 12 + 4 + 1 = 17 babies
  b 4 + 1 = 5 babies   **d** 2 + 8 + 12 + 4 + 1 = 27 babies

3 **a** $\frac{1}{4}$ **c** $\frac{1}{4}$ of 28 = 7 pupils
  b Any answer in range 40% to 48%

CHAPTER 5

1 **a** 2 m   **b** 10 m

2 **a** 20 cm   **b** 30 cm   **c** 40 cm   **d** 50 cm

3 **a** 900 g   **b** feet   **c** 15 cm   **d** pounds

4 **a** 8 + 12 + 6 + 10 = 36 cm   **b** 10 + 4 + 8 + 5 + 12 = 39 cm

CHAPTER 6

1 800 ÷ 100 = 8 glasses

2 **a** 6   **b** 23

3 **a** 5 cm^3 **b** 12 cm^3

4 **a** Volume = 4 × 3 × 6 = 72 cm^3
  b Volume = 20 × 5 × 37 = 3700 cm^3

CHAPTER 7

1 **a** 43.49 s
  b 43.49, 44.41, 44.53, 44.62, 44.70, 44.83, 44.99
  c 44.41 − 43.49 = 0.92
  d 44.99 − 43.49 = 1.5

2 **a** 3.67   4.27   4.29

    b 1.64   2.201   2.23

3 **a**

$$\begin{array}{r} 34.70 \\ + 21.20 \\ \hline 55.90 \end{array}$$ £55.90

    e

$$\begin{array}{r} 2.34 \\ \times \quad 3 \\ \hline 7.02 \\ \hline \end{array}$$ £7.02

    $\tiny 1 \ 1$

    b

$$\begin{array}{r} 4.80 \\ + 1.40 \\ \hline 6.20 \end{array}$$ £6.20

    f

$$\begin{array}{r} 12.45 \\ \times \quad 15 \\ \hline 124.50 \\ 62.25 \\ \hline 186.75 \end{array}$$ £186.75

    c

$$\begin{array}{r} 5.27 \\ - 4.15 \\ \hline 1.12 \end{array}$$ £1.12

    g $\begin{array}{r} 1.\,53 \\ 3\overline{)4.^159} \end{array}$ £1.53

    d

$\tiny ^2 \quad ^1$

$$\begin{array}{r} 2\not{3}.47 \\ - 12.73 \\ \hline 10.74 \end{array}$$ £10.74

    h $\begin{array}{r} 1.1\,5 \\ 5\overline{)5.7^25} \end{array}$ £1.15

4 **a** 3.29 ÷ 6

    [3] [.] [2] [9] [÷] [6] [=]   0.548333

    Answer: £0.55

    b Estimate: £3.29 is about £3

    £3 ÷ 6 = £0.50

    £0.50 is close to £0.55

5 Local school raises nearly £1000 for charity.

CHAPTER 8

1 **a**

16	Number of rows	Number of columns
	1	16
	2	8
	4	4
	8	2
	16	1

  c No

  f 1, 2, 4, 8, 16

2 **a** 198
    b Yes
    c 132
    d Yes
    e 79.2
    f No
    g 1, 4, 6, 9, 11, 12, 18, 22, 33, 36, 44, 66, 99, 132, 198

3 **a** $3 \times 3 = 9$ $4 \times 4 = 16$
    b 3, 19, 31
    c $8(= 2 \times 4)$ $16(= 4 \times 4)$

4

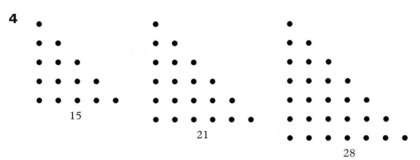

15   21   28

CHAPTER 9

1 **a** 4 minutes = 4×60 seconds
                = 240 seconds
    b 5 hours = 5×60 minutes
               = 300 minutes
    c 8 weeks = 8×7 days
              = 56 days
    d 4 years = 4×12 months
             = 48 months

2 1,   8,   15,   22,   29
       +7   +7   +7   +7

The rule is $+7$ as there are 7 days in a week.

3 **a** 3.10 pm   **b** 3.35 pm   **c** 12.15 am   **d** 6.55 am

4 **a** The pizza needs 25 minutes.
      At 5.42 pm it is 18 minutes to 6.00 pm.
      $25 - 18 = 7$
      The pizza is cooked at 6.07 pm.

b The stew needs 2 hours 15 minutes
It starts to cook at 3.50 pm.
Sort the 15 minutes out first.
3.50 pm + 15 minutes = 4.05 pm
The stew needs another 2 hours.
4.05 pm + 2 hours = 6.05 pm.
The stew is cooked at 6.05 pm.

5 a 10.23 am   **b** 2.25 pm   **c** 4.48 pm   **d** 10.47 pm

6 a 10.15 11.00 12.00 12.27

     45 mins 1 hour 27 mins

b 45 minutes + 1 hour + 27 minutes = 1 hour 72 minutes
There are 60 minutes in 1 hour so 72 minutes = 1 hour 12 minutes
The journey takes 2 hours 12 mins.

7 a 0815   **b** 0850   **c** 35 minutes
   d Paul will get the 0845 bus.
He waits from 0832 to 0845.
$45 - 32 = 13$
Paul waits 13 minutes.

CHAPTER 10

1 a $2°C > -7°C$ **b** $-12°C < -3°C$ **c** $-1°C < 8°C$

2 $-9°C, -5°C, 0°C, 4°C, 15°C$

3 $8°C - 12°C = -4°C$

4

Day temperature	Night temperature	Keys to press	Difference
5 °C	−6 °C	5 − 6 ± =	11°C
4 °C	−3 °C	4 − 3 ± =	7°C

5 a -2 **b** -4 **c** 2   **d** -4

6 $120°C - 15°C = 105°C$

7 a -1, next two terms are -4 and -5
   b -4, next two terms are 0 and -4

8 a 5 m   **b** 5 m   **c** 25 m   **d** 20 m

CHAPTER 11

1 a 175 + 180 + 185 + 176 + 184 = 900 cm
b 900 ÷ 5 = 180 cm

2 a 8
b 6
c 14 + 16 + 6 = 36
d 36 ÷ 3 = 12
e 7 + 5 + 3 = 15
   15 ÷ 3 = 5

3 a 0   3   4   ⑥   10   11   13
b 6 is the middle number. This is the median.

4 a £65 + £50 + £85 + £110 + £70 + £85 = £465
   £465 ÷ 6 = £77.50
b Biggest price = £110, smallest price = £50
   Range = 110 − 50 = £60

c £50   £65   │£70   £85│   £85   £110

   Median = $\dfrac{70 + 85}{2}$ = £77.50

   Put the prices in order. There are 2 middle prices. Find the mean of £70 and £85

CHAPTER 12

1 a

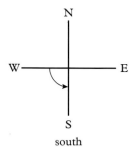

south

b

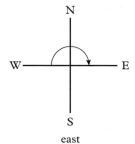

east

c

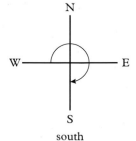

south

2 a The acute angles are 62°, 85°, 27°.
b The obtuse angles are 150°, 140°.
c The reflex angles are 225°, 290°, 305°.

3 a 90° + 70° = 160° but angles on a straight line always add up to 180°. The 70° must be wrong.
b The angle marked 120° is an acute angle. An acute angle is less than 90° so 120° is wrong.

c $120° + 100° + 60° = 280°$. The angles of a triangle always add up to 180° not 280°.

d $110° + 80° = 190°$ but angles on a straight line always add up to 180°. One angle must be wrong.

4

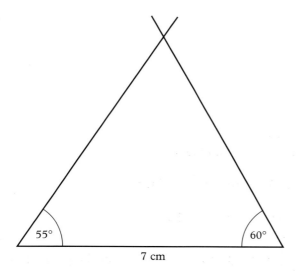

55°   60°

7 cm

CHAPTER 13

1 **a** $a + a + a = 3 \times a = 3a$
   b $g + g + g + g + g = 5 \times g = 5g$

2 **a** $m + m + m = 3m$
   b $3m - 12$

3 $2m + 4n$

4 **a**

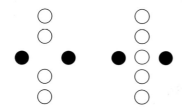

   b The 2 in the $n + 2$ rule stands for the black counters.
   c The n in the $n + 2$ rule stands for the white counters.
   d There are always 2 black counters.
   There will be $25 - 2 = 23$ white counters.

5   IN   OUT

$2 \longrightarrow 8 \longrightarrow 24 \longrightarrow 19$

$4 \longrightarrow 10 \longrightarrow 30 \longrightarrow 25$

$6 \longrightarrow 12 \longrightarrow 36 \longrightarrow 31$

$10 \longrightarrow 16 \longrightarrow 48 \longrightarrow 43$

CHAPTER 14

1   **a** $\frac{1}{2}$ of $20 = 20 \div 2 = 10$
  b $\frac{1}{4}$ of $20 = 20 \div 4 = 5$
  c $\frac{1}{4}$ of $20 = 5$ so $\frac{3}{4}$ of $20 = 3 \times 5 = 15$

2   They get 1 whole pizza and $\frac{1}{3}$ each.
The answer is $1\frac{1}{3}$

3   **a** $\frac{2}{8}$ **b** $\frac{12}{33}$

4 $\frac{8}{25} = 8 \div 25 = 0.32$

5 $\frac{2}{5} = 0.4$ $\frac{3}{10} = 0.3$ $\frac{2}{5}$ is larger

6 $\boxed{3}\ \boxed{\div}\ \boxed{4}\ \boxed{\times}\ \boxed{2}\ \boxed{8}\ \boxed{=}$ 21

7   **a** $\frac{2}{5}$
  b $\frac{2}{5}$ of $30 = 2 \div 5 \times 30 = 12$ chocolates

8   **a** £$20.40 \div 4 =$ £5.10
  b £$20.40 -$ £$5.10 =$ £15.30

CHAPTER 15

1   **a** Blue (because there are more blue cubes than red)
  b Red (because there are less red cubes than blue)
  c There are 2 red cubes. There are 6 cubes in total.
    Probability of getting a red $= \frac{2}{6}$
  d Probability of getting a blue $= \frac{4}{6}$ (there are 4 blue cubes)
  e Probability of getting a yellow $= \frac{0}{6} = 0$ because there are no yellow
    cubes.

2 **a** One red

   b $9 = 3 + 3 + 3$

   Rashid gets 1 red in each group of 3.

   He expects 3 reds.

   c $30 = 3 + 3 + 3 + 3 + 3 + 3 + 3 + 3 + 3 + 3$ i.e. 10×3

   He expects 10 reds.

3 **a** False. More pupils may choose one language than the other.
   It is not equally likely.

   b True

4 Survey (to find out how many pupils own a bicycle)

5

   b has been placed because there is a good chance that someone in the
   class has forgotten their exercise book.

CHAPTER 16

1 (1) **a** 14%   **b** $100\% - 14\% = 86\%$

   (2) **a** 72%   **b** $100\% - 72\% = 28\%$

2 $100\% - 33\% = 67\%$

3 **a** $10\% = \frac{10}{100} = \frac{1}{10}$ **c** $75\% = \frac{75}{100} = \frac{3}{4}$

   b $20\% = \frac{20}{100} = \frac{1}{5}$ **d** $70\% = \frac{70}{100} = \frac{7}{10}$

4 **a** There are 10 counters. There are 3 red counters.
   The fraction of counters that are red $= \frac{3}{10}$

   b $\frac{3}{10} = \frac{30}{100} = 30\%$ are red

   c There are 7 blue counters.
   The fraction of counters that are blue $= \frac{7}{10}$

   d $\frac{7}{10} = \frac{70}{100} = 70\%$ are blue

5 **a** $3\% = \frac{3}{100} = 3 \div 100 = 0.03$

   b $34\% = \frac{34}{100} = 34 \div 100 = 0.34$

   c $7\% = \frac{7}{100} = 7 \div 100 = 0.07$

   d $98\% = \frac{98}{100} = 98 \div 100 = 0.98$

6 a 15% of 60 = $\frac{15}{100} \times 60$

Keys to press: **1** **5** **÷** **1** **0** **0** **×** **6** **0** **=**

Answer 9 pupils.

b Number of pupils this year = 60 + 9

= 69

CHAPTER 17

1

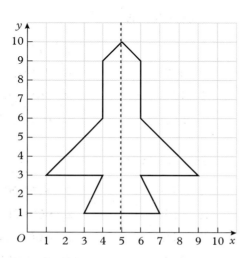

2 a (3, 2)   (6, 4)   (9, 6)

b The first co-ordinates are the 3× table.

The second co-ordinates are the 2× table.

c (12, 8)

(15, 10)

d 17 is not in the two times table.

Exercise 1

1	49	**5**	449	**9**	39
2	86	**6**	888	**10**	99
3	98	**7**	6949	**11**	646
4	896	**8**	8718	**12**	968

Exercise 2

1	41	**5**	301	**9**	84
2	62	**6**	531	**10**	149
3	65	**7**	441	**11**	213
4	182	**8**	845	**12**	390

Exercise 3

1	62	**5**	296	**9**	612
2	511	**6**	1512	**10**	521
3	222	**7**	8113	**11**	403
4	533	**8**	3324	**12**	1325

Exercise 4

1	18	**5**	104	**9**	369
2	46	**6**	569	**10**	474
3	27	**7**	149	**11**	218
4	218	**8**	204	**12**	179

Exercise 5

1	427	**3**	7715
2	353	**4**	1584

Exercise 6

1	66	**3**	124
2	396	**4**	399

Exercise 7

1	90	**6**	2140
2	70	**7**	1413
3	216	**8**	5706
4	375	**9**	2775
5	698	**10**	3604

Exercise 8

1	480	**5**	7770
2	540	**6**	24 960
3	1340	**7**	70 000
4	8420	**8**	90 030

Exercise 9

1	2700	**7**	415 300
2	9100	**8**	900 000
3	74 000	**9**	4 004 000
4	29 100	**10**	9 240 000
5	427 000	**11**	10 301
6	840 000	**12**	7 370 000

Exercise 10

1	780	**5**	1950	**9**	3550
2	840	**6**	1360	**10**	6080
3	780	**7**	3680	**11**	8730
4	690	**8**	1250	**12**	36 750

Exercise 11

1	675	**5**	9550	**9**	11 990
2	1824	**6**	14 634	**10**	53 365
3	6519	**7**	17 595	**11**	53 534
4	10 504	**8**	26 628	**12**	88 021

Exercise 12

1 5	**5** 9	**9** 8
2 11	**6** 4	**10** 9
3 4	**7** 4	**11** 8
4 4	**8** 7	**12** 9

Exercise 13

1 33	**4** 22
2 13	**5** 41
3 11	**6** 121

Exercise 14

1 27	**5** 13	**9** 108
2 17	**6** 17	**10** 142
3 16	**7** 15	**11** 177
4 13	**8** 18	**12** 54

Exercise 15

1 82	**5** 816
2 6	**6** 940
3 482	**7** 700
4 93	**8** 50 000

Exercise 16

1 54	**5** 840
2 71	**6** 84
3 82	**7** 400
4 64	**8** 40

Exercise 17

1 41	**5** 376
2 16	**6** 154
3 93	**7** 22
4 105	**8** 306

Exercise 18

1 $\frac{3}{8}$ **5** $\frac{1}{2}$ **9** $\frac{5}{24}$

2 $\frac{1}{5}$ **6** $\frac{7}{12}$ **10** $\frac{5}{12}$

3 $\frac{1}{13}$ **7** $\frac{1}{12}$

4 $\frac{3}{10}$ **8** $\frac{1}{6}$

Exercise 19

1 **a** 34 mm **c** 1310 mm
   b 128 mm **d** 1137 mm

2 **a** 410 cm **c** 1210 cm
   b 280 cm **d** 32 400 cm

3 **a** 8800 m **c** 15 000 m
   b 9700 m **d** 100 000 m

Exercise 20

1 **a** 3000 g **d** 3 kg
   b 7000 g **e** 6 kg
   c 21 000 g **f** 0.8 kg

2 **a** 3500 g **d** 6.5 kg
   b 7200 g **e** 3.2 kg
   c 3840 g **f** 2.8 kg

3 **a** 5000 mg **d** 3 t
   b 8 g **e** 4000 kg
   c 1.64 g **f** 2500 kg